Advances in Geological Science

Studies in the twentieth century uncovered groundbreaking facts in geophysics and produced a radically new picture of the Earth's history. However, in some respects it also created more puzzles for the research community of the twenty-first century to tackle. This book series aims to present the state of the art of contemporary geological studies and offers the opportunity to discuss major open problems in geosciences and their phenomena. The main focus is on physical geological features such as geomorphology, petrology, sedimentology, geotectonics, volcanology, seismology, glaciology, and their environmental impacts. The monographs in the series, including multi-authored volumes, will examine prominent features of past events up to their current status, and possibly forecast some aspects of the foreseeable future. The guiding principle is that understanding the fundamentals and applied methodology of overlapping fields will be key to paving the way for the next generation.

Ming-Ching Tom Kuo

Groundwater Radon in the Taiwan Subduction Zone

A Natural Strain-Meter for Earthquake Prediction

Springer

Ming-Ching Tom Kuo
Department of Resources Engineering
National Cheng Kung University
Tainan, Taiwan

ISSN 2524-3829 ISSN 2524-3837 (electronic)
Advances in Geological Science
ISBN 978-981-99-5349-3 ISBN 978-981-99-5350-9 (eBook)
https://doi.org/10.1007/978-981-99-5350-9

This Springer imprint is published by the registered company Springer Nature Singapore Pte Ltd.
The registered company address is: 152 Beach Road, #21-01/04 Gateway East, Singapore 189721, Singapore

Preface

This book was written to provide practical information on monitoring anomalous declines in groundwater-radon concentration for the early warning of local disastrous earthquakes, based on the author's real-life experience in the Taiwan subduction zone. Most published radon anomalies show increases in the concentration of groundwater radon. Very little information is currently available for field data manifesting anomalous declines in groundwater-radon concentration. The scope of this book focuses on anomalous declines in groundwater-radon concentration precursory to earthquakes. The data on anomalous declines in groundwater-radon concentration are valuable. The anomaly can be applied to quantify the gas saturation and volumetric strain in a low-porosity aquifer through simple physical models.

The radon anomaly observed in Taiwan before the 2003 Mw 6.8 Chengkung earthquake corroborates the anomalous decrease in groundwater-radon concentration recorded in Japan before the 1978 Izu-Oshima-Kinkai earthquake of magnitude 7.0. Both lucky observations reveal that the groundwater radon, when observed under suitable geological conditions, can be a sensitive tracer for strain changes in the crust preceding an earthquake. Nonetheless, what are plausible mechanisms to explain the radon anomalies observed before the Taiwan 2003 Chengkung and the Japan 1978 Izu-Oshima-Kinkai earthquakes?

The above question has been motivating my research since 2003 to focus on anomalous declines in groundwater-radon concentration precursory to earthquakes. With the help of case studies in Taiwan subduction zone, this book presents:

1. a mechanism of in-situ radon volatilization for interpreting anomalous decrease in groundwater radon,
2. suitable geological conditions to site a well which detects anomalous declines in groundwater-radon concentration consistently,
3. a well-test method to confirm that the observed anomalous decrease in groundwater radon is induced by in-situ radon volatilization.

A small low-porosity fractured aquifer near an active fault can be an effective natural strainmeter for earthquake warnings. A suitable geological site in the Taiwan subduction zone cited in this book is Antung hot spring. Recurrences of anomalous declines in the concentration of groundwater radon were consistently recorded at Antung hot spring precursory to four main earthquakes near the Longitudinal Valley Fault (the 2003 Mw 6.8 Chengkung, 2006 Mw 6.1 and Mw 5.9 Taitung, and 2008 Mw 5.4 Antung earthquakes).

Well tests provide field evidence for the in-situ development of gas bubbles in water-saturated fractured rock prior to the 2008 Mw 5.4 Antung earthquake. Anomalous declines in groundwater radon and aquifer transmissivity are two precursory phenomena having a common effect of gas-bubble development. This is the first discovery of aquifer transmissivity complementing groundwater radon as an earthquake precursor. The findings provide a complementary method supporting crustal deformation as the cause of observed anomalous decrease in groundwater radon, not groundwater mixing. The learnings in this book can be applied globally in the subduction zone with similar tectonic settings and physical–chemical relationships.

I owe my own development to associations with many people of Stanford Geothermal Program. My mentors at Stanford University, Paul Kruger and W. E. Brigham, set me on this path with skills in radon measurements and well tests. Both skills have been essential to my groundwater-radon research since 2003 in the Taiwan subduction zone. I have also been a beneficiary of groundwater-radon research for earthquake prediction in Japan. For this I particularly would like to thank F. Tsunomori, H. Wakita, M. Noguchi, and S. Tasaka.

Many people have helped in my preparation of this book. In particular I acknowledge the assistance of my editor, Yosuke Nishida, who offered many helpful comments on the manuscript. Support from National Cheng Kung University and NCKU Research and Development Foundation, Tainan, Taiwan is appreciated. The author is grateful to W. S. Chen for providing the geological map of the Antung area, and to Judy Kuo for drawing the schematic of the in-situ radon-volatilization mechanism. I dedicate the book to my wife, Mei, and my daughters, Betty and Judy, who encouraged me to be good and strong to carry out this task.

Tainan, Taiwan Ming-Ching Tom Kuo

Acknowledgments

I would like to thank the following copyright holders who have graciously permitted the reproduction of material in this book.

SPRINGER NATURE
From *Pure and Applied Geophysics*: Fig. 7.11, Kuo et al. (2017) 174: 1291–1301; Figs. 7.6–7.10, Kuo et al. (2020) 177: 2877–2887.
From *Hydrogeology Journal*: Figs. 2.1, 2.4–2.6, 2.9–2.11, Han et al. (2006) 14: 173–179.
From *Natural Hazards*: Figs. 5.1 and 5.2, Kuo et al. (2011) 59: 861–869.

Stanford University
From *Stanford Geothermal Program*: Figs. 2.3, 2.7, and 2.8 Stoker and Kruger (1975) SGPTR4: 1–116.

AMERICAN GEOPHYSICAL UNION
From *Journal of Geophysical Research*: Figs. 6.7 and 6.8, Sugisaki and Sugiura (1986) 91: 12296–12304; Figs. 2.2, 7.3–7.5, Hauksson and Goddard (1981) 86: 7037–7054.

Elsevier
From *Applied Radiation and Isotopes*: Figs. 6.1–6.5, Kuo et al. (2018) 136: 68–72.
From *Radiation Measurements:* Figs. 7.1 and 7.2, Tsunomori and Kuo (2010) 45: 139–142.
From *Journal of Environmental Radioactivity*: Figs. 3.1 and 3.2, Kuo et al. (2006a) 88: 101–106.
I would also like to thank Dr. Denis E. Bergeron, Editor-in-Chief of *Applied Radiation and Isotopes* and Dr. Lenny Konikow, Editor-in-Chief of *Groundwater* for their encouragement and permission to reuse and extend the following articles to the book. From *Applied Radiation and Isotopes:* Kuo et al. (2018) 136: 68–72.
From *Groundwater*: Kuo et al. (2006) 44: 642–647; Kuo et al. (2022) 60: 510–517.

Contents

About the Author

Ming-Ching Tom Kuo Ph.D. is a professor of petroleum engineering at National Cheng Kung University, Taiwan. He gained his industrial experience while working at Stanford Research Institute, Occidental Petroleum, and Mobil. He is experienced in oil, gas, and geothermal production. His research interests include groundwater hydrology, groundwater and soil remediation, and monitoring of groundwater radon. Since 2003, his research has focused on the applications of recurrent radon precursors to forecast local large earthquakes in the Taiwan subduction zone.

Dr. Kuo earned his B.S. degree in chemical engineering from National Taiwan University, an M.S. in chemical engineering from Kansas State University, and a Ph.D. in environmental engineering from Stanford University.

List of Figures

List of Tables

Chapter 1
Introduction

Abstract The data of anomalous declines in groundwater-radon concentration are valuable for earthquake prediction. A small low-porosity brittle aquifer near an active fault can be applied as an effective natural strainmeter to detect anomalous declines in groundwater-radon concentration consistently. To quantify volumetric strain in aquifer rock, it is essential to confirm crustal deformation as the cause of observed anomalous decrease in groundwater radon, not groundwater mixing. This introductory chapter will summarize the objectives and contents of the book.

Keywords Radon · Groundwater · Earthquake · Natural strainmeter

1.1 Background and Objectives

The purpose of this book is to provide practical information on monitoring anomalous declines in groundwater-radon concentration for the early warning of local disastrous earthquakes, based on the author's real-life experience in the Taiwan subduction zone. Most published radon anomalies show increases in the concentration of groundwater radon. Very little information is currently available for field data manifesting anomalous declines in groundwater-radon concentration. The scope of this book focuses on anomalous declines in groundwater-radon concentration precursory to earthquakes. The data on anomalous declines in groundwater-radon concentration are valuable. The anomaly can be applied to quantify the gas saturation and volumetric strain in a low-porosity aquifer through simple physical models (Kuo et al. 2006). Worldwide case studies from Taiwan, Japan, the United States, and Iceland are reviewed and compared to each other. With the help of case studies in Taiwan subduction zone, this book presents the application of recurrent anomalous declines in groundwater radon for forecasting local large earthquakes.

The following chapters present:

1. suitable geological conditions to site a well which detects anomalous declines in groundwater-radon concentration consistently,

M. T. Kuo, *Groundwater Radon in the Taiwan Subduction Zone*, Advances in Geological Science,
https://doi.org/10.1007/978-981-99-5350-9_1

2. a mechanism of in-situ radon volatilization for interpreting anomalous decrease in groundwater radon,
3. a well-test method to confirm that the observed anomalous decrease in groundwater radon is induced by in-situ radon volatilization.

A small low-porosity fractured aquifer can be an effective natural strainmeter for earthquake warnings. When regional stress increases, micro-cracks in brittle rock could form at a faster rate than the rate of groundwater recharge in the aquifer under undrained conditions. As a result, gas bubbles develop and induce detectable earthquake precursors. A suitable geological site in the Taiwan subduction zone cited in this book is Antung andesite spring. Recurrences of anomalous declines in the concentration of groundwater radon were consistently recorded at Antung well D1 precursory to four main earthquakes near the Longitudinal Valley Fault (the 2003 M_w 6.8 Chengkung, 2006 M_w 6.1 and M_w 5.9 Taitung, and 2008 M_w 5.4 Antung earthquakes). Via a basic observation of groundwater radon at Antung well D1, all large thrust-type earthquakes occurring on the Longitudinal Valley Fault can be warned months in advance.

Well tests provide field evidence for the in-situ development of gas bubbles in water-saturated fractured rock prior to the 2008 M_w 5.4 Antung earthquake (Kuo 2022). Anomalous declines in groundwater radon and aquifer transmissivity are two precursory phenomena having a common effect on gas-bubble development. This is the first discovery of aquifer transmissivity complementing groundwater radon as an earthquake precursor. The findings provide a complementary method supporting crustal deformation as the cause of observed anomalous decrease in groundwater radon, not groundwater mixing. The learnings in this book can be applied globally in the subduction zone with similar tectonic settings and physical–chemical relationships.

1.2 Organization of the Book

In addition to this introductory chapter, the book is divided into the following six chapters:

Chapter 2: Methods of Monitoring Groundwater Radon. This chapter illustrates methods of monitoring groundwater radon including groundwater sampling and radon determination. Both methods of discrete sampling and continuous monitoring are reviewed. Radon distributions in groundwater near an active fault are discussed. The application of groundwater radon for earthquake prediction research is also briefly reviewed.

Chapter 3: Anomalous Radon Decline at Antung Hot Spring before the 2003 Mw 6.8 Chengkung Earthquake. The 2003 Chengkung earthquake was the strongest earthquake near the Chengkung area since 1951. This chapter presents the radon anomaly observed at the Antung radon-monitoring well D1 located 20 km from the epicenter. The 2003 radon anomaly observed at the Antung hot spring corroborates the radon anomaly recorded before the 1978 Izu-Oshima-Kinkai earthquake. Both observations at Antung and Izu suggest that the groundwater radon, when monitored at suitable geological sites, can be a sensitive tracer for strain changes in the crust preceding an earthquake. Both anomalies motivated further research to answer the following questions: (1) a physical mechanism, or, in-situ radon-volatilization mechanism to interpret anomalous decrease in groundwater-radon precursory to an earthquake, (2) recurrences of radon anomalies precursory to local main earthquakes, (3) confirmation of in-situ radon-volatilization mechanism, and (4) suitable geological conditions to site a well for monitoring anomalous declines in groundwater-radon concentration.

Chapter 4: A Physical Mechanism of Groundwater-Radon Volatilization. Plausible mechanisms for interpreting anomalous decreases in groundwater-radon concentration prior to earthquakes are seldom discussed in the literature. This chapter presents a physical mechanism to explain the anomalous decrease in groundwater-radon concentration before the Taiwan M_w 6.8 Chengkung earthquake. With the help of the geology at Antung spring, this chapter also outlines suitable geological conditions to site groundwater radon-monitoring wells for catching radon anomalies. Like groundwater-dissolved radon, other groundwater-dissolved gas components, such as methane, ethane, etc., also volatilize from groundwater into gas bubbles. The in-situ volatilization mechanism extends naturally from dissolved radon to other dissolved gas components.

Chapter 5: Recurrences of Radon Anomalies Precursory to Local Main Earthquakes at Antung Hot Spring. This chapter presents recurrent anomalous declines in groundwater-radon concentration consistently recorded at Antung prior to local main earthquakes that occurred between 2003 and 2010. Recurrences of radon anomalies are essential in earthquake prediction. The formation of gas bubbles is necessary for the phenomena of in-situ radon volatilization to occur in a fractured aquifer. Well tests provide field evidence regarding the in-situ development of gas bubbles in water-saturated fractured rock prior to the 2008 M_w 5.4 Antung earthquake. Anomalous declines in groundwater radon and aquifer transmissivity are two precursory phenomena having a common effect of gas bubbles. For the first time, it is discovered that aquifer transmissivity can complement groundwater radon as an earthquake precursor.

Chapter 6: Anomalous Radon Declines in Small Unconfined Aquifers: Corroboration of Favorable Geological Conditions. Suitable geological conditions at Antung hot spring (a low-porosity brittle aquifer surrounded by a ductile formation in undrained conditions) are attributed to catch the recurrent radon anomalous declines precursory to the local main large earthquakes. The above hypothesis can be tested by monitoring groundwater radon at the Paihe Spring with similar geological conditions at Antung. This chapter presents the radon anomalous declines recorded at the Paihe Spring prior to the 2010 M_w 6.3 Jiasian and 2016 M_w 6.4 Meinong earthquakes. The other hard-rock site discussed in this chapter is Byakko Spring in Japan. Precursory to the 1984 M 6.8 Western Nagano earthquake, anomalous declines in groundwater-dissolved gas were observed at the Byakko Spring. The 1984 anomaly recorded at the

Byakko Spring not only corroborated the mechanism of in-situ volatilization but also confirmed the favorable geological conditions for catching gas anomalous declines.

Chapter 7: Anomalous Radon Declines in Small Confined Aquifers: Evaluation of Global Data. Worldwide data manifesting anomalous declines in groundwater-radon concentration are reviewed. Data from Japan, Iceland, and Taiwan are evaluated and compared each other. The data are evaluated in terms of geological conditions, monitoring methods, epicenter distance, radon-decline percent, precursory time, and earthquake magnitude. With the help of a case study at the Antung hot spring, this chapter also illustrates the application of recurrent radon precursors for forecasting local large earthquakes occurring on a given active fault (Longitudinal Valley Fault) in the Taiwan subduction zone.

References

Kuo MCT et al (2006) A mechanism for anomalous decline in radon precursory to an earthquake. Ground Water 44(5):642–647

Kuo MCT (2022) Study of water relative permeability in fractures using well tests and radon: gas bubbles effect. Ground Water 60(4):510–517

Chapter 2
Methods of Monitoring Groundwater Radon

Abstract In this chapter, we will present methods of monitoring groundwater radon including groundwater sampling and radon determination. Radon distributions in groundwater near an active fault are discussed. The application of groundwater radon for earthquake prediction research is also briefly reviewed.

Keywords Radon-222 · Fault · Groundwater · Earthquake-prediction research

2.1 Introduction

Measurement of radon-222 in groundwater has been often applied as a possible precursor in earthquake prediction studies (Kuo et al. 2006; Wakita et al. 1980; Wakita 1996; Shapiro et al. 1980; Kuo et al. 2013; Noguchi and Wakita 1977; Hauksson and Goddard 1981; Hauksson 1981; Roeloffs 1999; Trique et al. 1999; Erees et al. 2007; Kuo 2014; Tarakçı et al. 2014; Tsunomori and Tanaka 2014; Kuo et al. 2017, 2018; Morales-Simfors et al. 2020). Based on a worldwide survey, most anomalies precursory to earthquakes revealed increases in groundwater-radon concentration while a few anomalies manifested decreases in groundwater-radon concentration (Hauksson 1981). Very little information is currently available for field data manifesting anomalous declines in groundwater-radon concentration. This book focuses on the anomalous declines in groundwater-radon precursory to main earthquakes in Taiwan (Kuo et al. 2006, 2013), Japan (Wakita et al. 1980; Wakita 1996), and Iceland (Hauksson and Goddard 1981).

In this chapter, methods of monitoring groundwater radon will be addressed. Case studies from Taiwan, Japan, and Iceland are presented to illustrate procedures of sample collection and radon determination. Discrete samples of groundwater samples were practiced in case studies of Taiwan and Iceland. The sampling frequency was about twice per week in both Taiwan and Iceland. At Izu Peninsula in Japan, continuous monitoring of groundwater radon was applied. Groundwater samples collected in Taiwan are in a single liquid phase while geothermal-water samples collected in

M. T. Kuo, *Groundwater Radon in the Taiwan Subduction Zone*, Advances in Geological Science,
https://doi.org/10.1007/978-981-99-5350-9_2

Iceland are in two phases (vapor and liquid). Case studies from Taiwan, Japan, and Iceland are reviewed to illustrate the monitoring of groundwater radon for earthquake prediction research.

2.2 Collection of Groundwater Samples

The most common and basic sampling procedure for groundwater radon is collecting single-phase groundwater samples. The collection of single-phase groundwater samples will be presented with case studies in Taiwan. For geothermal water, suitable sampling methods are required for two-phase samples. The collection of two-phase samples of geothermal water will be presented with case studies from Iceland and the United States.

2.2.1 Single-Phase Groundwater Samples

To obtain representative groundwater samples for radon measurements of appropriately constructed wells are essential. Radon concentration in groundwater is related to the emanation rates of geological layers. Representative sampling must be from appropriate monitoring wells which consist of suitable bentonite seals. The monitoring wells enable vertically spaced water samples to be taken through screens and filter gravel.

Inadequate purging can be a major source of error, because the water sample is a mixture of stagnant water from the well bore, pore water from the filter gravel, and groundwater (Freyer et al. 1997; Han et al. 2006). Radon concentration in groundwater is influenced by the natural emanation rate of the aquifer. A submersible pump is often used for groundwater-radon sampling in monitoring wells. Every sampling starts with flushing the stagnant water in the well. Figure 2.1 shows an example of the radon concentration in the well discharge during continuous sampling (Han et al. 2006). During the first period of flushing, the radon concentration of the water samples is practically zero and then increases rapidly to the mean radon concentration for this monitoring well, 529 ± 19 pCi/L. Figure 2.1 also indicates that a purging of 3 well-bore volumes is necessary and adequate before taking samples for radon measurements.

A 40-ml glass vial with a TEFLON lined cap is used for sample collection. Radon tends to escape from groundwater to the headspace of the sample vial. It is essential that the sample vial is inverted to check for air bubbles each time right after collecting a sample. If any bubbles are present, the sample is discarded and the sampling procedure repeats. The date and time of sampling is recorded and the sample is stored in a cooler. The maximum holding time before radon counting is 4 days.

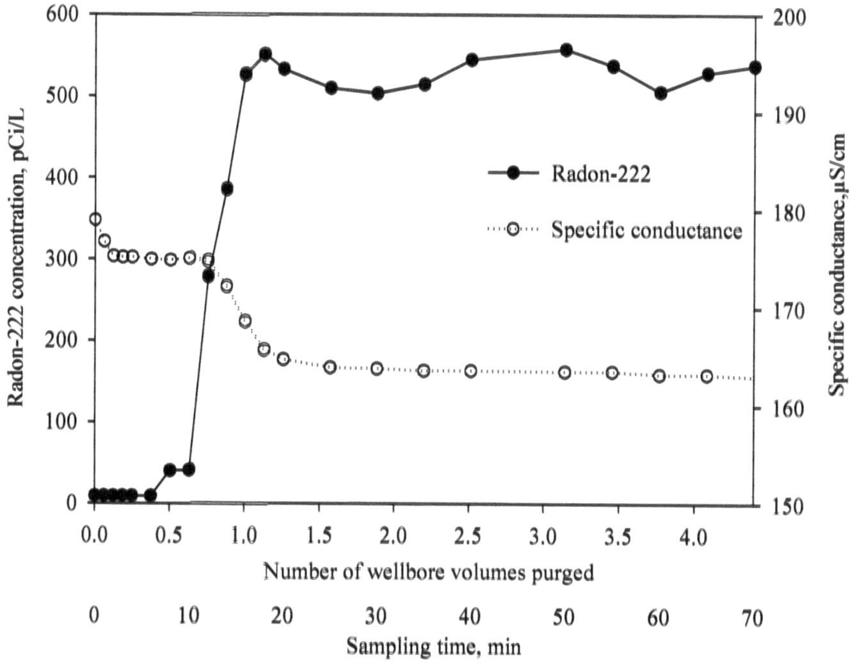

Fig. 2.1 Radon concentration and electrical conductivity in the well discharge during continuous sampling in a deep observation well (From Han et al. 2006)

2.2.2 *Two-Phase Samples of Geo-Fluids*

When the geothermal water produces through a well, dissolved gases flash from water as bubbles. A precise way of sampling two-phase geothermal water is to draw a total sample. Figure 2.2 shows the apparatus used in Iceland to sample geothermal wells (Hauksson and Goddard 1981). An evacuated bottle with a vacuum gauge is connected to the wellhead during sampling. Then, fill the sample bottle until the vacuum gauge reaches 1 atm pressure. Close the clamps on the hoses of the sample bottle before disconnecting the sample bottle from the sampling apparatus. When a gas sample is taken, the sampling bottle is first filled with well water. Then, the sample bottle is inverted to allow gas to expel water from the sample bottle.

Stoker and Kruger (1975) developed standard procedures for sampling geothermal wells. Figure 2.3 shows the sampling apparatus connected to the wellhead. Steam and hot water samples are obtained in a 4.7-L evacuated steel cylinder. A "T" fitting with a valve and bleed line allows purging the connecting line prior to sampling. The evacuated bottles were used to obtain reproducible samples of radon in hot water from liquid-dominated geothermal reservoirs and in steam condensate from vapor-dominated geothermal reservoirs.

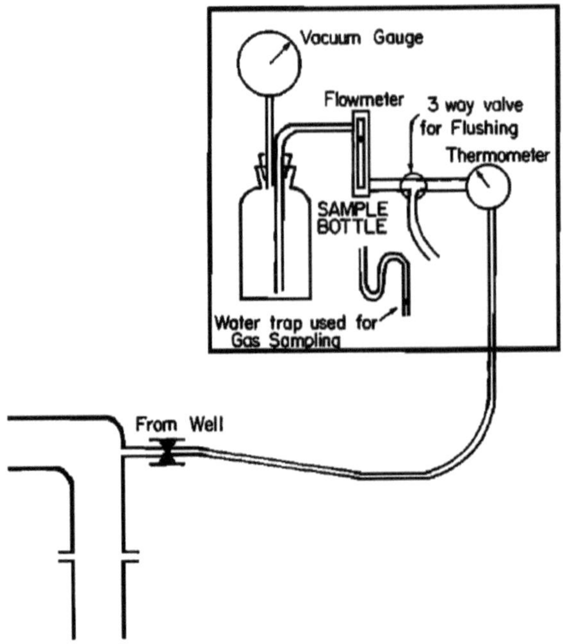

Fig. 2.2 Apparatus used in Iceland to sample two-phase samples from low-temperature geothermal wells (From Hauksson and Goddard 1981)

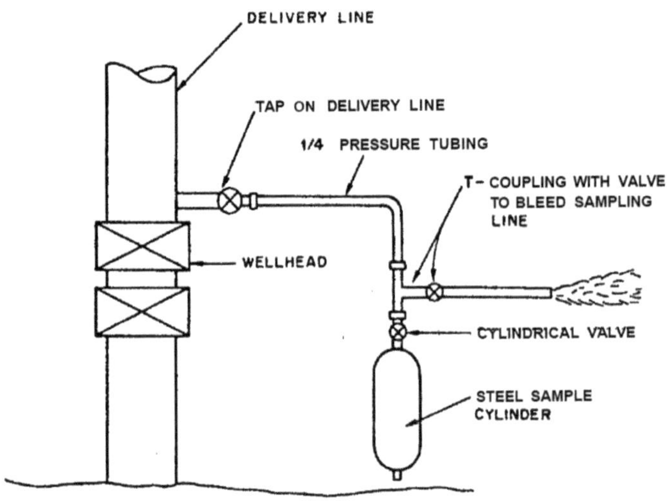

Fig. 2.3 Apparatus used in the United States to sample two-phase samples from high-temperature geothermal wells (From Stoker and Kruger 1975)

2.3 Determination of Radon Activity

Radon in groundwater samples can be measured by several counting methods based on detecting alpha particles from radon and its daughters. This section presents two common methods of radon measurement: liquid scintillation counting method and ZnS scintillation counting method. In the liquid scintillation counting method, a mineral-oil scintillation cocktail is used to extract radon from groundwater samples. In the ZnS scintillation counting method, purge and trap is applied to extract radon from a groundwater sample into an evacuated counting cell (Lucas cell).

2.3.1 Liquid Scintillation Counting Method

Radon is highly soluble in organic liquids such as toluene or mineral oil. Radon in groundwater samples can be measured by liquid scintillation counting method (Noguchi 1964; Prichard and Gesell 1977). The liquid scintillation method has the advantages of minimal sample preparation time (about 1 min/sample), small sample size (10–15 ml), automatic sample changing, and a good detection limit (below 20 pCi/L for a 50-min count using the sample volume of 15 ml).

The minimal sample preparation time required by liquid scintillation counting method makes a large number of groundwater samples that can be readily processed for radon measurement. Radon concentrations in groundwater are determined by drawing a 15-ml sample directly from a field sample into a clean syringe. The samples are injected beneath a 5-ml layer of mineral oil-based scintillation solution in 24-ml counting vials to prevent the loss of radon from the samples in the process. Radon is partitioned selectively into a mineral-oil scintillation cocktail immiscible with the water sample. The vials are vigorously shaken to promote phase contact and the extraction efficiency of radon from water phase into oil phase. The 24-ml counting vials are dark-adapted and equilibrated for about 3 h after extraction (Prichard and Gesell 1977) and then assayed with a liquid scintillation counter. The results are corrected for the amount of radon decay between sampling and assay.

The decay chain from radon-222 through polonium-214 involves the release of three alphas (5.490, 6.003, and 7.687 meV). Figure 2.4 shows an example of the alpha spectrum using the TRI-CARB software of the Packard 1600TR counter. The peaks of radon-222 (5.490 meV), polonium-218 (6.003 meV), and polonium-214 (7.687 meV) can be distinguished. Counting with an optimized window (Lowry 1991), a detection limit of 10 pCi/L can be achieved for a 40-min count using the sample volume of 10 ml (Prichard et al. 1992).

The results of the measurements from the liquid scintillation counter are determined in units of counts per minute (cpm). Using an aqueous Ra-226 calibration solution, which is in secular equilibrium with Rn-222 progen, an average calibration factor can be obtained for unit conversion from cpm to pCi. Figure 2.5 shows an example of calibration factor of 7.1 ± 0.1 cpm/pCi and background < 6 cpm for

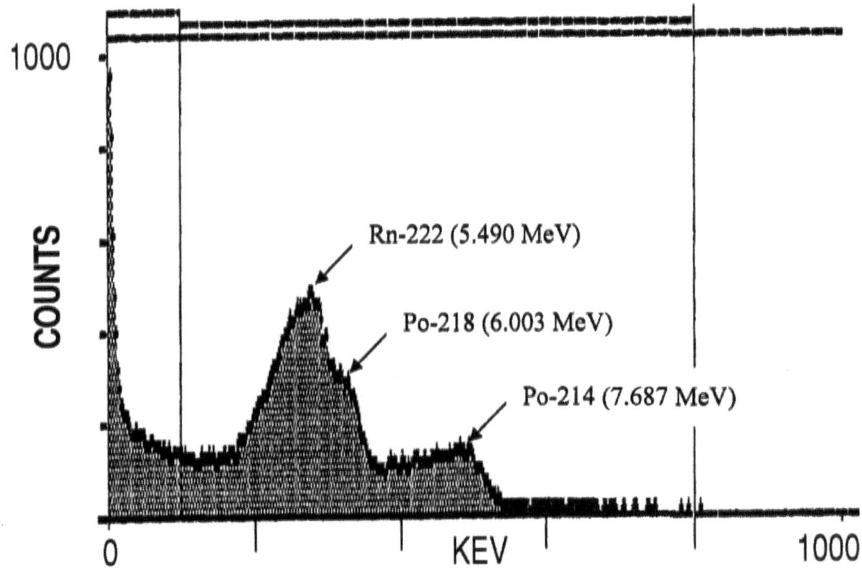

Fig. 2.4 Alpha spectrum of radon-222 and its daughter nuclides represented by TRI-CARB software (From Han et al. 2006)

liquid scintillation measurements. With a count time of 50 min and background < 6 cpm, a detection limit below 20 pCi/L can be achieved using the sample volume of 15 ml.

Figure 2.6 shows the half-life measurement of radon decay by repeated counting of groundwater samples from two wells in Taiwan. Figure 2.6 shows that the half-life of 3.84 days experimentally determined from a semi-log straight line using samples from Well Liu-Ying (I) agrees well with the literature value of 3.83 days. The half-life of 3.56 days experimentally determined for samples from Well Wen-Tsu (II) is a little shorter than 3.83 days. Lack of tightness at the caps of counting vials could cause radon loss and introduce an error in the measurement of radon half-life. Figure 2.6 provides an effective means to verify that radon-222 is the main radioisotope responsible for radioactivity in the well water.

2.3.2 ZnS Scintillation Counting Method

Lucas (1957, 1964) developed a low-background ZnS scintillation counter to survey both Radon-222 and Radium-226. ZnS scintillation counting method was for the measurement of radon content in geothermal water (Stoker and Kruger 1975; Hauksson and Goddard 1981) and in groundwater (Noguchi and Wakita 1977; Tsunomori and Tanaka 2014).

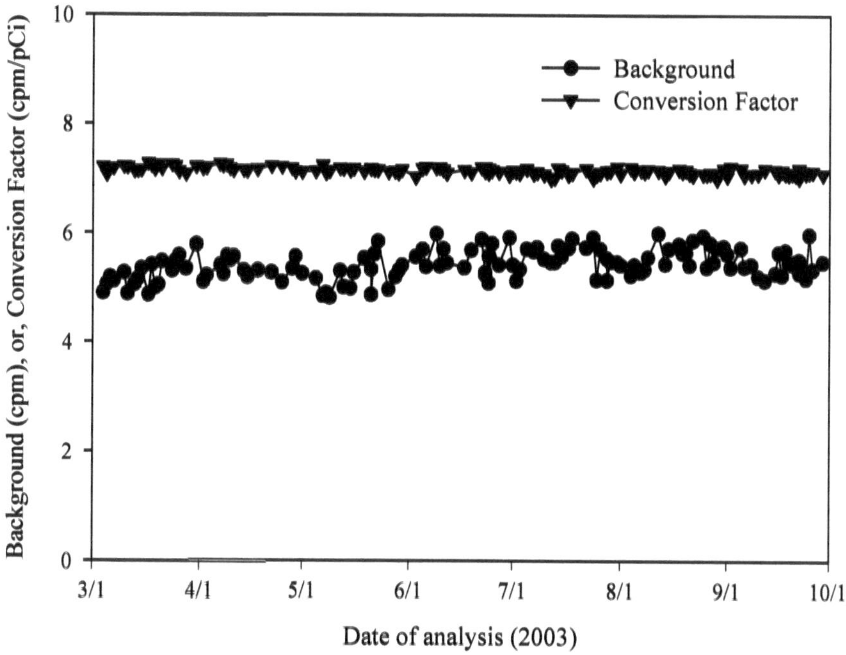

Fig. 2.5 Calibration factor and background for LSC measurements (From Han et al. 2006)

Figure 2.7 shows a purge and trap system used by Stoker and Kruger (1975) to purge radon out of the samples with helium carrier gas. The extraction process consists of the adsorption of radon on activated carbon at −80 °C, followed by thermal desorption at 400 °C, and transfer in helium counting gas into a low-background ZnS scintillation counting cell (Lucas cell). Radon is adsorbed quantitatively on activated charcoal in the carbon adsorption trap for radon at −80 °C in a Freon TA-dry ice bath. Then, the dry ice bath is removed and an aliquot of helium counting gas from the He reservoir is let into the trap. Radon is desorbed from the activated charcoal at 400 °C by a heating element bonded onto the glass. The desorbed radon is transferred with successive aliquots of helium counting gas through a peristaltic pump into an evacuated ZnS scintillation counting cell (Lucas cell). Hauksson and Goddard (1981) also applied similar techniques to study radon earthquake precursors in Iceland.

Figure 2.8 shows the schematic diagram of the radon detection system using the ZnS scintillation counting method (Stoker and Kruger 1975). The system consists of a Lucas scintillation flask optically coupled to a photomultiplier tube in a light-tight housing. Lucas cell is the detector with the ZnS phosphor coating on the inside of its cylindrical surface. Alpha particles from radon-222 and its daughters (Polonium-218 and Polonium-214) produce light pulses in the ZnS phosphor. Light pulses pass through the quartz window which forms the bottom of Lucas scintillation flask. The alpha particle-generated pulses are counted in a singlechannel pulse-height analyzer set at a threshold to discriminate noise pulses.

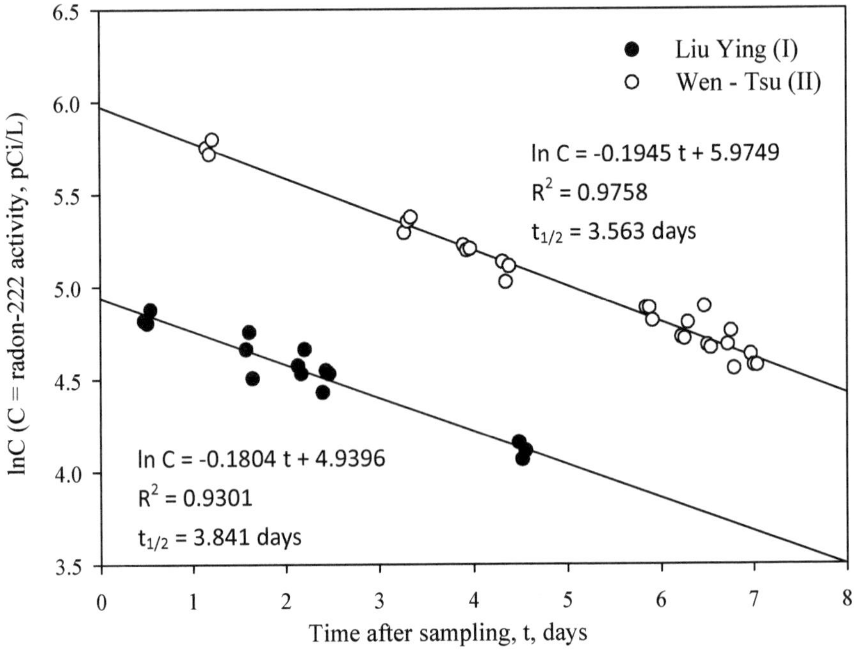

Fig. 2.6 Measurement of half-life from semi-logarithmic decay curve (From Han et al. 2006)

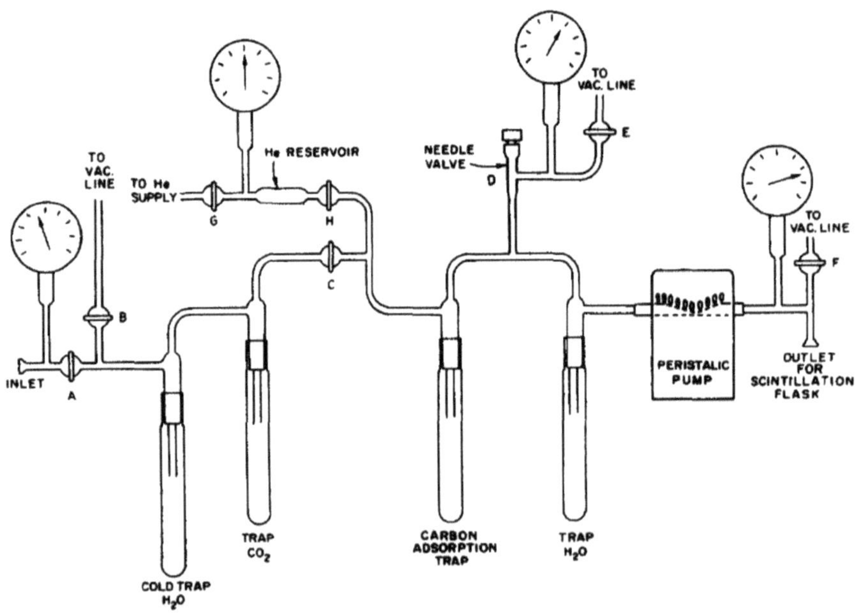

Fig. 2.7 Schematic diagram of the purge and trap system for extracting radon from geo-fluids (From Stoker and Kruger 1975)

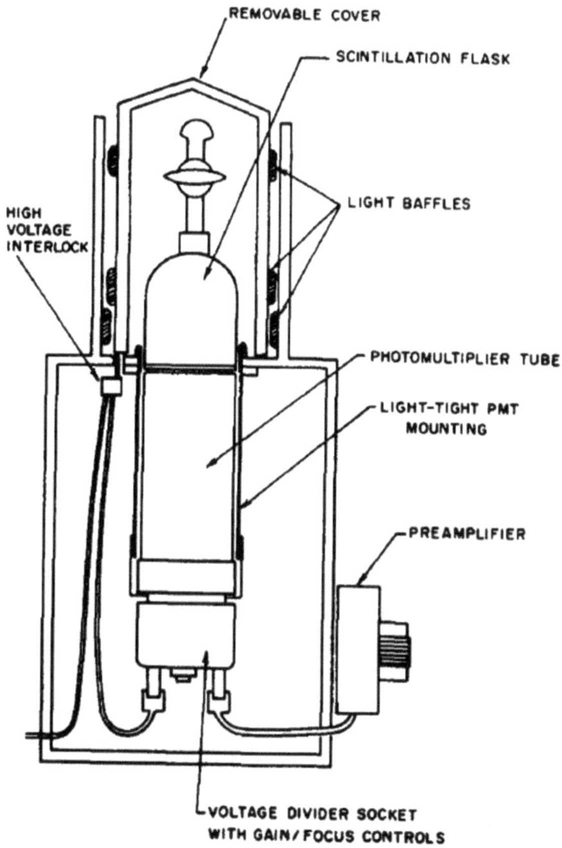

Fig. 2.8 Schematic diagram of the radon detection system using Lucas scintillation flask (From Stoker and Kruger 1975)

2.3.3 Continuous Measurement of Radon in Groundwater

ZnS scintillation counter is considered to be the most suitable detector for continuous measurement of radon in groundwater for earthquake prediction studies in Japan (Noguchi and Wakita 1977; Wakita et al. 1980). The continuous measuring system consists of a separation chamber in which radon volatilizes from groundwater to the gas phase and a ZnS scintillation chamber for alpha counting. The radon concentration in the gas phase is in equilibrium with that in the groundwater according to Henry's constant. Alpha particles from radon-222 and its daughters (Polonium-218 and Polonium-214) in the scintillation chamber are counted and recorded continuously. The counting rate is about 100 cpm/100 pCi Rn/L for the continuous measuring system (Noguchi and Wakita 1977).

Tsunomori and Tanaka (2014) also built a continuous radon measuring system at SKE-1 well near Nakaizu on the Izu Peninsula, Japan. A counter-flow air-stripping extractor was used to separate radon efficiently from groundwater to the gas phase for radon counting by the ZnS scintillation method (Tsunomori and Tanaka 2014).

Air bubbles flow counter-currently through flowing water extracting radon from groundwater to the air bubbles. Their continuous measuring system at SKE-1 well can provide a long-term monitoring of groundwater radon at low operating and maintenance budget.

In a continuous radon measuring system, two alternative methods are used to extract the dissolved radon from groundwater to air: water spraying and diffusion through the membrane (Vyletělová and Froňka 2019). In the water-spraying method, nozzles are used in a separation unit to spray the groundwater extracting radon from the droplets of water to the air. In the membrane method, the waterproof membrane usually is a several meters long hose which is immersed in groundwater and air flows inside the hose. Radon diffuses from groundwater through the membrane to the circulating air.

2.4 Radon Distributions in Groundwater Near an Active Fault

Radon levels were surveyed in the Pingtung Plain, which is bounded by Taiwan Strait to the west and by the Chaochou fault to the east. Figure 2.9 shows the name, location, and extent of the Pingtung Plain and other major groundwater areas in Taiwan. Yearly total groundwater consumption in the Pingtung Plain is about 31% of the total groundwater consumption in Taiwan.

Figure 2.10 shows the distribution of radon-222 in groundwater in the Pingtung Plain. The radon-222 concentration in groundwater varied in a wide range from below the detection limit of 18 pCi/L up to 1,100 pCi/L in the Pingtung Plain. However, the wells with radon concentrations higher than 400 pCi/L are all located near the Chaochou fault. Anomalous high concentrations of groundwater radon were observed along the active Chaochou fault (Han et al. 2006).

Figure 2.11 shows the concentration distribution in groundwater radon along an east–west cross section AA' in the Pingtung Plain. Four well clusters are on the cross section AA' (W, H, T, and C) starting from the Chaochou fault on the east side to Taiwan Strait on the west side. The lithology of well cluster W is mainly gravel more than 200 m thick and is located in the area of groundwater recharge. Figure 2.11 illustrates that the radon concentration near the Chaochou fault in the deep aquifer is higher than that in the shallow aquifer. At well cluster W in the recharge area, the radon concentrations in the deep and shallow aquifers were 528 ± 43 pCi/l and 276 ± 6 pCi/l, respectively. The radon concentrations in groundwater decrease as the wells move away from the Chaochou fault toward Taiwan Strait.

Anomalous high concentrations of groundwater radon were also observed to coincide with geological fault systems in Texas, United States (Cech et al. 1988). In Japan, Tsunomori et al. (2017) surveyed radon concentrations in groundwater around the Tachikawa fault zone. The radon concentrations in bedrock groundwater samples

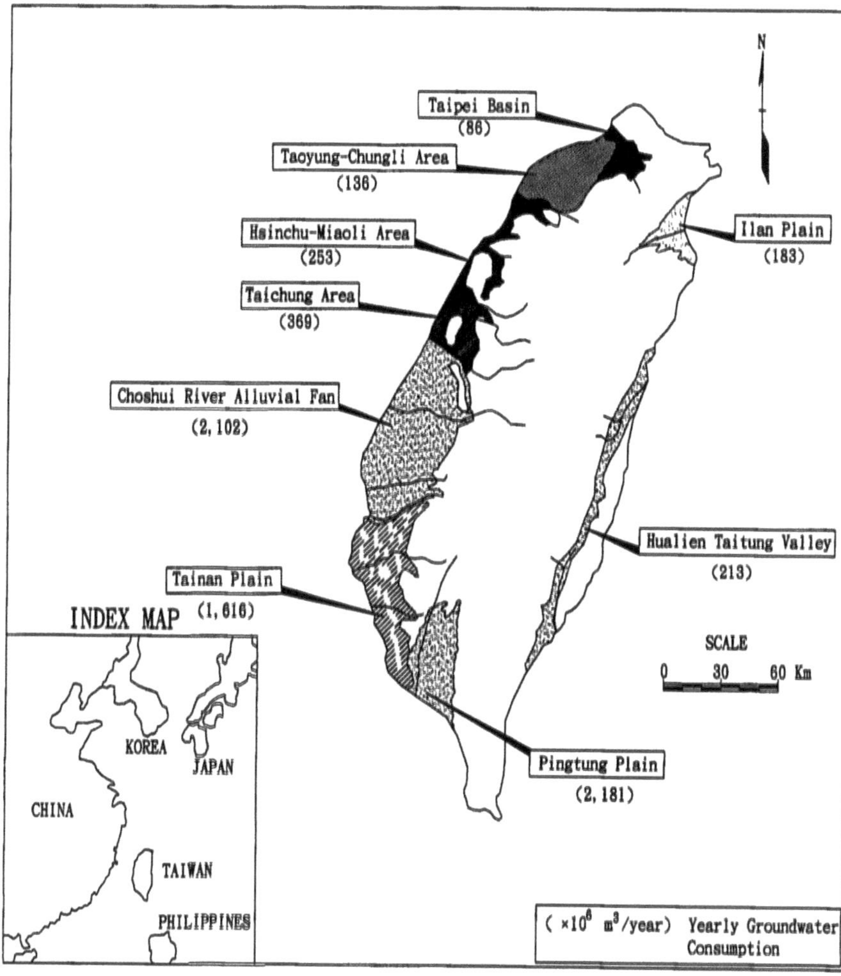

Fig. 2.9 Major groundwater areas in Taiwan and yearly groundwater consumption (From Han et al. 2006)

beside the Naguri and Tachikawa fault segments are significantly higher than the radon concentrations expected from the geology on the Kanto plane.

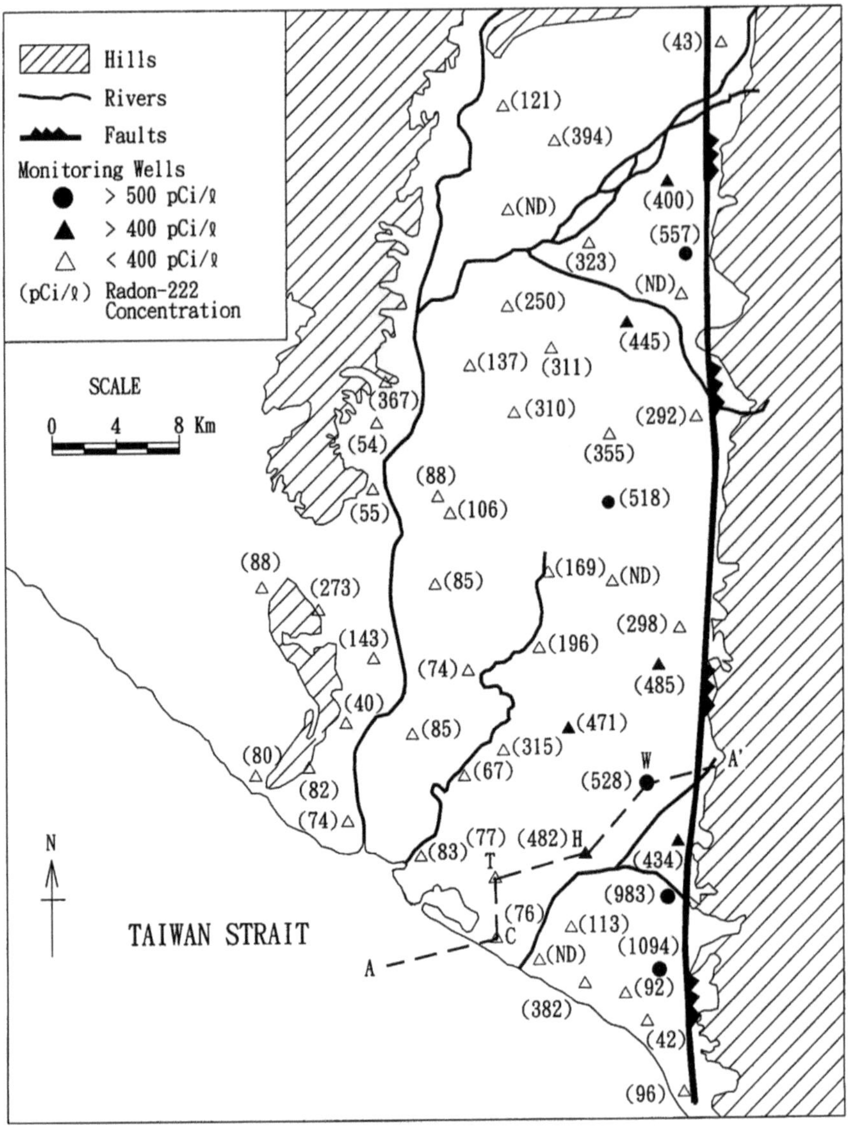

Fig. 2.10 Distribution of radon-222 in groundwater in the Pingtung Plain (From Han et al. 2006)

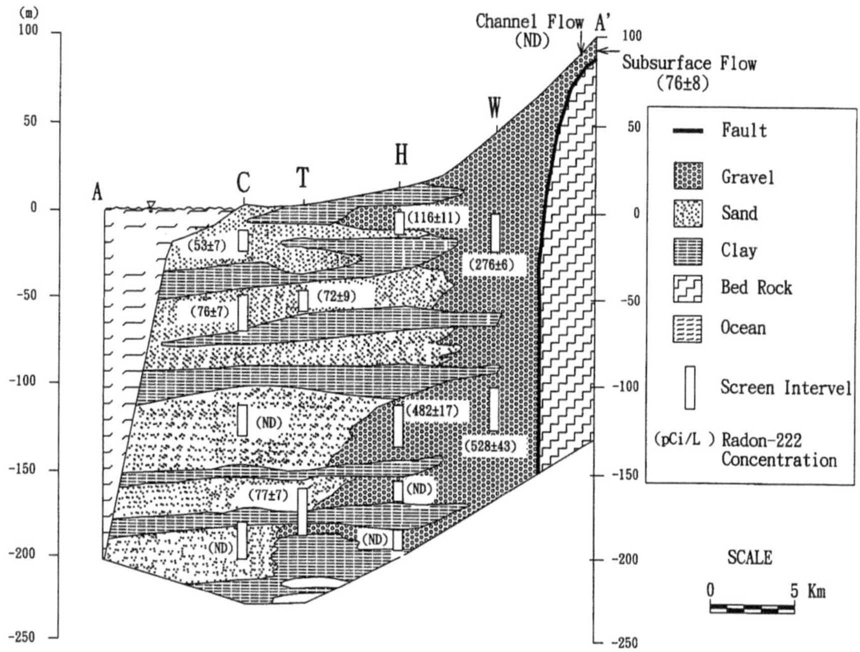

Fig. 2.11 Distribution of radon-222 in groundwater with the distance from the Chaochou fault along geological cross section AA' in the Pingtung Plain (From Han et al. 2006)

2.5 Application of Groundwater Radon for Earthquake Prediction Research

Radon-222, a radioactive nuclide with a half-life of 3.83 days, is chemically inert and highly soluble in water. The transport behavior of radon in geological environments is governed by physical processes such as fluid advection, diffusion, mass transfer of radon between liquid and gas phases, and radioactive decay. The radon concentration in groundwater is proportional to the uranium content in adjacent rocks in an aquifer (Andrews and Wood 1972). Groundwater-radon concentration can be related to the average radium-226 activity of the aquifer matrix as follows (Andrews and Wood 1972).

$$C_W^{Ra} = \frac{\varepsilon A_{Ra} \rho_b}{n} \tag{2.1}$$

where C_W^{Ra} is the equilibrium radon concentration in groundwater with 100% water saturation, pCi/L; A_{Ra} is the radium activity concentration of the aquifer matrix, pCi/kg; n is the porosity of the aquifer matrix, fraction; ρ_b is the bulk density of the aquifer matrix, kg/L; and ε is the dimensionless emanation power which quantifies the fraction of the produced radon atoms emanate into the pore space. Because of radon's short recoil length (3 *10^{-6} cm), only atoms produced near the surface of rock grains are released to the surrounding groundwater (Tanner 1964). The radon concentration in groundwater is mainly dependent on the surface area of rocks (Torgersen et al. 1990). Before the occurrence of an earthquake, when regional stress increases, the formation of micro-cracks in rock masses could cause an increase in the surface area of rocks, and radon concentration often rises in the groundwater. For example, Igarashi et al. (1995) reported anomalous increases in the concentration of groundwater-radon precursory to the 1995 Kobe earthquake in Japan. The change in groundwater-radon concentration precursory to the Kobe earthquake may reflect the strain in aquifer rocks. Mass transfer of radon between liquid and gas phases in aquifers is an important mechanism causing anomalous declines in the concentration of groundwater-radon precursory to large local earthquakes (Kuo et al. 2006). The partition of radon between water and gas phases in aquifer is governed by Henry's law and constant (Weigel 1978). In-situ radon-volatilization model was established to interpret the observed anomalous radon-declines precursory to local earthquakes (Kuo et al. 2006, 2013, 2017, 2018).

Monitoring of radon-222 in groundwater has been applied for earthquake prediction (Kuo et al. 2006; Wakita et al. 1980; Wakita 1996; Shapiro et al. 1980; Kuo et al. 2013; Noguchi and Wakita 1977; Hauksson and Goddard 1981; Hauksson 1981; Roeloffs 1999; Trique et al. 1999; Erees et al. 2007; Kuo 2014; Tarakçı et al. 2014; Tsunomori and Tanaka 2014; Kuo et al. 2017; Kuo et al. 2018; Morales-Simfors et al. 2020). Most radon anomalies showed an increase in the radon concentration of groundwater according to a worldwide survey (Hauksson 1981). Contrarily, few anomalies manifested a decrease in the radon content of groundwater (Kuo et al. 2006; Wakita et al. 1980; Wakita 1996; Shapiro et al. 1980; Hauksson and Goddard 1981). The anomalous declines in groundwater radon reported in the literature are all observed in low-porosity undrained brittle aquifers. This book focuses on anomalous declines in groundwater-radon-concentration precursory to local main earthquakes. Simple analytical models will be presented in the book to study the precursory changes in gas saturation and volumetric strain in low-porosity brittle aquifers with undrained conditions.

Anomalous declines in the concentration of groundwater were observed to precede both the 1978 M 7.0 Izu-Oshima-Kinkai earthquake in Japan (Wakita et al. 1980; Wakita 1996) and the 2003 M_w 6.8 Chengkung earthquake in Taiwan (Kuo et al. 2006). Both observations suggest that the groundwater radon, when observed at suitable sites, can be a sensitive tracer for strain changes in the crust associated with earthquake occurrences. Details of the anomalous declines in groundwater radon recorded before the 2003 M_w 6.8 Chengkung earthquake will be presented in the next chapter and compared to the 1978 M 7.0 Izu-Oshima-Kinkai earthquake.

References

Andrews JN, Wood DF (1972) Mechanism of radon release in rock matrices and entry into groundwater. Trans Inst Min Metall B81:198–209

Cech I et al (1988) Radon distribution in domestic water of Texas. Ground Water 26(5):561–569

Erees FS et al (2007) Radon concentrations in thermal waters related to seismic events along faults in the Denizli Basin, Western Turkey. Radiat Meas 42:80–86

Freyer K et al (1997) Sampling and measurement of radon-222 in water. J Environ Radioact 37(3):327–337

Han YL et al (2006) Radon distribution in groundwater of Taiwan. Hydrogeol J 14:173–179

Hauksson E (1981) Radon content of groundwater as an earthquake precursor: evaluation of worldwide data and physical basis. J Geophys Res: Solid Earth 86(B10):9397–9410

Hauksson E, Goddard JG (1981) Radon earthquake precursor studies in Iceland. J Geophys Res: Solid Earth 86(B8):7037–7054

Igarashi G et al (1995) Ground-water radon anomaly before the Kobe earthquake in Japan. Science 269:60–61

Kuo MCT et al (2006) A mechanism for anomalous decline in radon precursory to an earthquake. Ground Water 44(5):642–647

Kuo T (2014) Correlating precursory declines in groundwater radon with earthquake magnitude. Ground Water 52(2):217–224

Kuo T, Chen W, Ho C (2018) Anomalous decrease in groundwater radon before 2016 Mw 6.4 Meinong earthquake and its application in Taiwan. Appl Radiat Isot 136:68–72

Kuo T et al (2013) Concurrent concentration declines in groundwater-dissolved radon, methane and ethane precursory to 2011 MW 5.0 Chimei earthquake. Radiat Meas 58:121–127

Kuo T et al (2017) A stress condition in aquifer rock for detecting anomalous radon decline precursory to an earthquake. Pure Appl Geophys 174:1291–1301

Lowry JD (1991) Measuring low radon levels in drinking water supplies. J Am Water Works Assoc 83(4):149–153

Lucas HF (1957) Improved low-level alpha-scintillation counter for radon. Rev Sci Instrum 28(9):680–683

Lucas HF (1964) A fast and accurate survey technique for both Radon-222 and Radium-226. In: Adams JAS, Lowder WM (eds) The natural radiation environment. University of Chicago Press, Chicago, IL, pp 315–329

Morales-Simfors N, Wyss RA, Bundschuh J (2020) Recent progress in radon-based monitoring as seismic and volcanic precursor: A critical review. Crit Rev Environ Sci Technol 50(10):979–1012

Noguchi M (1964) Radioactivity measurement of radon by means of liquid scintillation fluid. Radioisotope 13(5):362–366

Noguchi M, Wakita H (1977) A method for continuous measurement of radon in groundwater for earthquake prediction. J Geophys Res 82:1353–1357

Prichard HM, Gesell TF (1977) Rapid measurements of 222Rn concentrations in water with a commercial liquid scintillation counter. Health Phys 33:577–581

Prichard HM, Venso EA, Dodson CL (1992) Liquid-Scintillation analysis of 222Rn in water by alpha-beta discrimination. Journal of Radioactivity and Radiochemistry 3(1):28–36

Roeloffs E (1999) Radon and rock deformation. Nature 399:104–105

Shapiro MH, Melvin JD, Tombrello TA (1980) Automated radon monitoring at a hard-rock site in the southern California transverse ranges. J Geophys Res 85:3058–3064

Stoker A, Kruger P (1975) Radon measurements in geothermal systems. SGP-TR-4. Stanford University, Stanford, CA.

Tanner AB (1964) Radon migration in the ground: a review. In: Adams JAS, Lowder WM (eds) The natural radiation environment. University of Chicago Press, Chicago, pp 161–196

Tarakçı M et al (2014) Investigation of the relationships between seismic activities and radon level in western Turkey. Appl Radiat Isot 83:12–17

Torgersen T, Benoit J, Mackie D (1990) Controls on groundwater Rn-222 concentrations in fractured rock. Geophys Res Lett 17(6):845–848

Trique M et al (1999) Radon emanation and electric potential variations associated with transient deformation near reservoir lakes. Nature 399:137–141

Tsunomori F, Tanaka H (2014) Anomalous change of groundwater radon concentration monitored at Nakaizu well in 2011. Radiat Meas 60:35–41

Tsunomori F et al (2017) Radon concentration distributions in shallow and deep groundwater around the Tachikawa fault zone. J Environ Radioact 172:106–112

Vyletělová P, Froňka A (2019) Continuous radon-in-water monitoring-comparison of methods under laboratory conditions and results of in situ measurements. Radiat Prot Dosim 186(2–3):406–412

Wakita H (1996) Geochemical challenge to earthquake prediction. Proc Natl Acad Sci USA 93:3781–3786

Wakita H et al (1980) Radon anomaly: a possible precursor of the 1978 Izu-Oshima-kinkai earthquake. Science 207(4433):882–883

Weigel F (1978) Radon. Chemiker-Zeitung 102:287–299

Chapter 3
Anomalous Radon Decline at Antung Hot Spring Before the 2003 Mw 6.8 Chengkung Earthquake

Abstract In this chapter, we will present the anomalous radon decline observed at Antung hot spring before the 2003 M_w 6.8 Chengkung Earthquake, which was the strongest earthquake on the Longitudinal Valley Fault since 1951. The 2003 radon anomaly observed in Taiwan corroborates the radon anomaly recorded before the 1978 Izu-Oshima-Kinkai earthquake in Japan. Both observations at Antung and Izu suggest that the groundwater radon, when monitored at suitable geological sites, can be a sensitive tracer for strain changes in the crust preceding an earthquake.

Keywords Radon anomaly · Fault · Groundwater · Earthquake precursor

3.1 Introduction

An earthquake of magnitude (M_w 6.8) occurred on December 10, 2003, near the Chengkung area in eastern Taiwan, which is the strongest since 1951. Precursory changes in the groundwater-radon concentration were observed at the Antung radon-monitoring station located 24 km from the epicenter. Approximately 65 days prior to the 2003 M_w 6.8 Chengkung earthquake, the radon anomaly was a decrease from a background level of 787 pCi/L to a minimum of 326 pCi/L. Observations at the Antung hot spring corroborate the anomalous declines in groundwater-radon concentration recorded before the 1978 Izu-Oshima-Kinkai earthquake of magnitude 7.0 (Wakita et al. 1980). Both observations suggest that the groundwater radon, when observed under suitable geological conditions, can be a sensitive tracer for strain changes in the crust preceding an earthquake.

M. T. Kuo, *Groundwater Radon in the Taiwan Subduction Zone*, Advances in Geological Science,
https://doi.org/10.1007/978-981-99-5350-9_3

3.2 Radon Monitoring at Antung Near the Longitudinal Valley Fault

Figure 3.1 illustrates Taiwan's tectonic setting and location of the Longitudinal Valley Fault which is the present-day plate suture between the Eurasian and the Philippine Sea plates. The Chihshang Fault is the most active segment of the Longitudinal Valley Fault which ruptured during two 1951 earthquakes of magnitude M 6.2 and M 7.0 (Hsu 1962). Based on the annual survey of geodetic and GPS measurements, the active creeping Chihshang Fault has been moving at a rapid steady rate, about 2–3 cm yr^{-1} (Angelier et al. 2000; Yu & Kuo 2001).

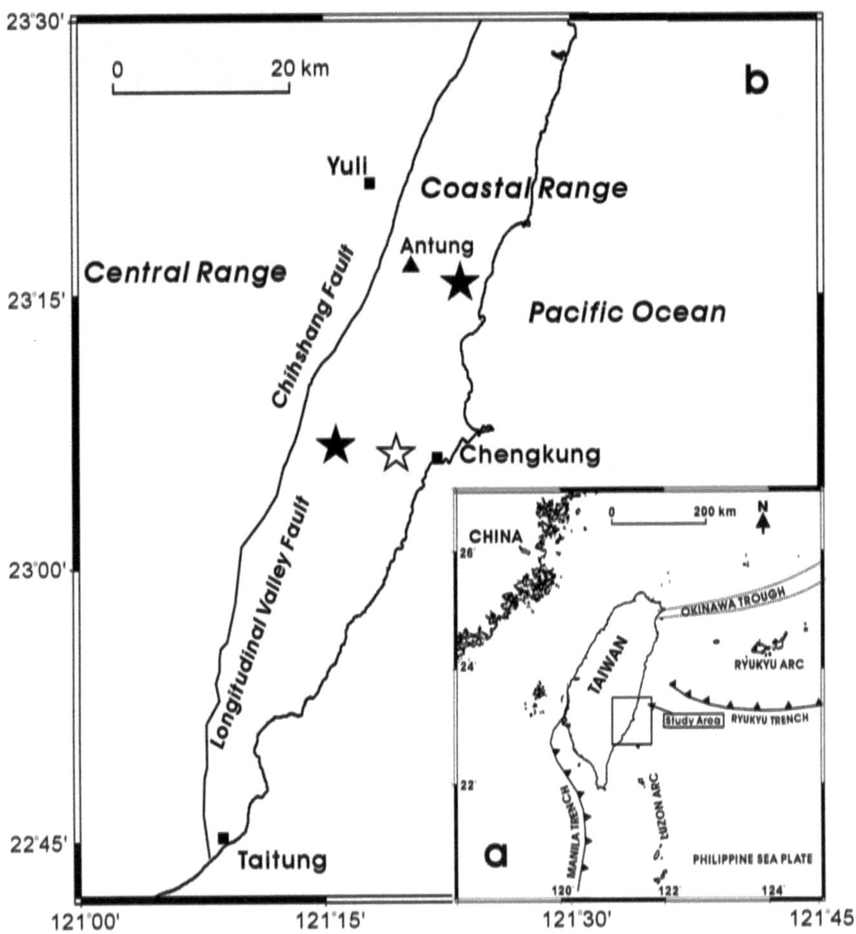

Fig. 3.1 (a) Tectonic setting of Taiwan (study area: location of b). (b) Location map of the Longitudinal Valley Fault area. The open star is the 2003 mainshock; filled stars are the 1951 mainshocks; filled triangle is radon-monitoring station (From Kuo et al. 2006)

Figure 3.2 shows the surface slip history of the Chihshang Fault from mid-1986 to December 2004 (Lee et al. 2005). The dashed line represents direct measurements of offsets during the period 1986–1997 (Angelier et al. 1997, 2000). The average creeping rate was 27 mm yr^{-1} and 22 mm yr^{-1} in 1986–1992 and in 1992–1997, respectively. The solid line represents creepmeter monitoring during the period 1998–2003 (Lee et al. 2005). The average creeping rate was 17 mm yr^{-1} during the period 2000–2003. The decreasing creeping rate since 2000 suggests increasing seismic hazards along the Chihshang Fault.

The Antung hot spring located only about 3 km southeast of the Chihshang Fault (or, the Longitudinal Valley Fault) was selected for the radon-monitoring site. The filled triangle in Fig. 3.1 shows the location of the radon-monitoring well D1. Monitoring of groundwater radon started in July 2003 at well D1 located in the Antung hot spring. The monitoring well D1 is located 24 km north of the hypocenter of the 2003 M_w 6.8 Chengkung earthquake.

Discrete samples of geothermal water have been collected from the monitoring well D1 for analysis of radon (Rn-222) content. The sampling frequency was about

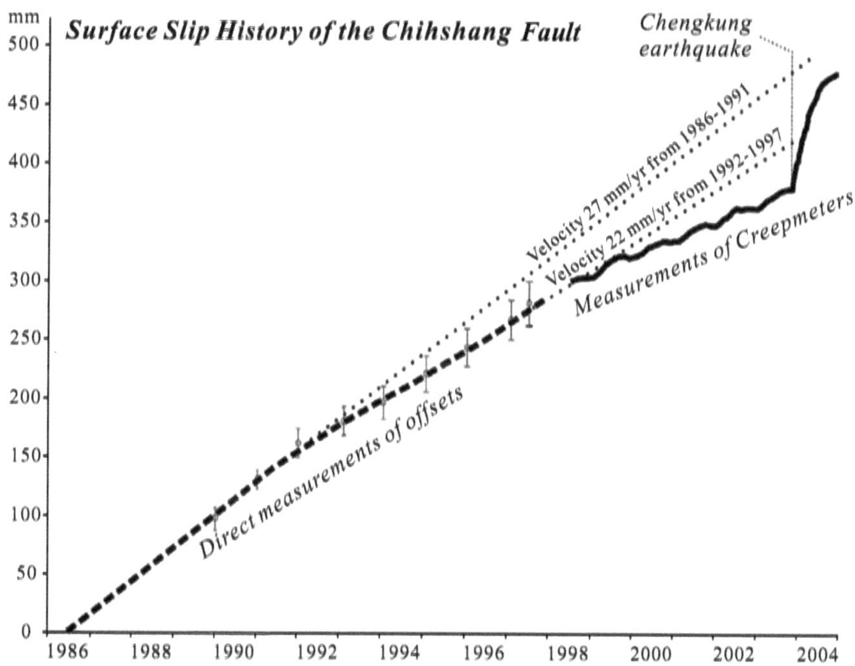

Fig. 3.2 Shortening across the Chihshang Fault, from mid-1986 to December 2004, with a tentative estimate of strain deficit before the Chengkung earthquake (December 10, 2003) (From Lee et al., 2005). Shortening (mm) versus time (years). Dashed line: results 1986–1997. Solid line: results 1998–2003, Chinyuan creepmeters. Uncertainties as error bars (for creepmeter data, within curve thickness). Dotted lines: extrapolation of aseismic creep shortening until the Chengkung earthquake: upper bound from 1986 to 1991, lower bound from 1992 to 1997, respectively, giving minimum strain deficits of 106 mm and 46 mm in December 2003 (From Kuo et al. 2006)

twice per week. The production interval of the Antung well D1 ranges from 167 to 187 m below the ground surface. The well was pumped more or less continuously.

A submersible pump is used for groundwater-radon sampling in the monitoring well D1. Every sampling required preliminary flushing with stagnant water in the monitoring well. Before obtaining samples for radon measurements, a minimum of three well-bored volumes were purged. An insufficiently purged volume introduces a major source of error—the water sample would contain a mixture of stagnant water from the monitoring well, pore water from the filter gravel, and groundwater emitted by the natural emanation rate of the aquifer.

During the sampling procedure, transportation, and preparation, prevention of radon escaping is important. A 40-ml glass vial with a TEFLON lined cap was used for sample collection at Antung well D1. Groundwater-dissolved radon tends to escape from groundwater to the headspace and gas bubbles of the sample vial. After collecting a sample, the sample vial was inverted to check for air bubbles. If any bubbles were present in the vial, the sample water was discarded and sampling was repeated. The date and time of sample collection were recorded. Then the samples were stored and transported in a cooler. Counting radioactivity was completed within 4 days.

The liquid scintillation counting method described by Noguchi (1964) was adopted to determine the concentration of radon-222 in groundwater samples from well D1. Radon concentrations are determined by drawing a 15-ml sample directly from a field sample into a clean syringe. The samples are injected beneath a 5-ml layer of mineral oil-based scintillation solution in 24-ml counting vials to prevent aeration of the samples in the process. The vials are vigorously shaken to promote phase contact and the extraction efficiency of radon from water phase into oil phase. The 24-ml counting vials are then assayed with a liquid scintillation counter.

An average calibration factor of 7.1 ± 0.1 cpm/pCi was obtained by using an aqueous Ra-226 calibration solution. Using a count time of 50 min and a sample volume of 15 ml, a detection limit below 18 pCi/L was achieved with a background of less than 6 cpm.

3.3 Radon Anomaly Before the 2003 M_w 6.8 Chengkung Earthquake

Figure 3.3 shows the sequence of events for radon data prior to the 2003 M_w 6.8 Chengkung earthquake. The radon concentration in groundwater was fairly stable (787 pCi/L on average) in the period from July 2003 to September 2003. Sixty-five days before the 2003 M_w 6.8 Chengkung earthquake which occurred on December 10, 2003, the radon concentration in groundwater started to decrease for 45 days. Twenty days prior to the earthquake, the radon concentration reached a minimum value of 326 pCi/L and then started to increase again. The radon concentration recovered to the background level (787 pCi/L on average) before the 2003 Chengkung earthquake.

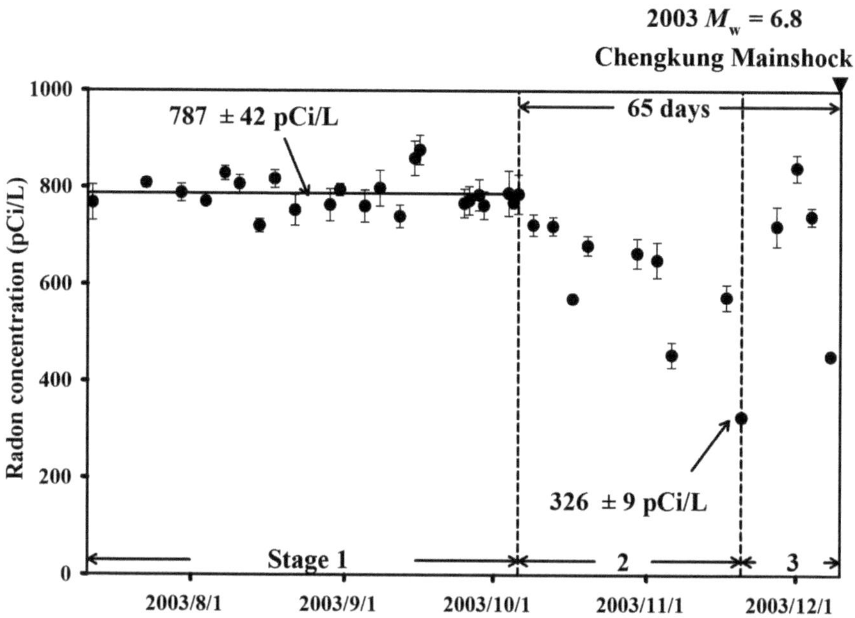

Fig. 3.3 Radon-concentration data prior to the 2003 M_W 6.8 Chengkung earthquake at the monitoring well (D1) in the Antung hot spring (From Kuo et al. 2006)

Radon-concentration errors shown in Fig. 3.3 are ± 1 standard deviation after simple averaging of triplicates. Based on the data trend as shown in Fig. 3.3, the radon data of the above anomaly consist of three stages. During stage 1 (from July 10, 2003 to October 5, 2003), the radon concentration in groundwater was fairly stable. During stage 2 (from October 5, 2003 to November 20, 2003), the trend of radon concentration in groundwater was decreasing. During stage 3 (from November 20, 2003 to December 10, 2003), the trend of radon concentration in groundwater was increasing.

Environmental records such as temperature and rainfall were examined to make sure that the radon anomaly was not attributed to these environmental factors. The aquifer temperature was at 60 °C. There was no heavy rainfall during the observation period responsible for the radon anomaly.

3.4 Criticism of Radon as a Precursor

The radon anomaly observed before the 2003 M_W 6.8 Chengkung earthquake corroborates the anomalous decrease in groundwater-radon concentration recorded before the 1978 Izu-Oshima-Kinkai earthquake of magnitude 7.0 (Wakita et al. 1980).

Both observations reveal that the groundwater radon, when observed under suitable geological conditions, can be a sensitive tracer for strain changes in the crust preceding an earthquake. Nonetheless, what are plausible mechanisms to explain the radon anomalies observed before the 2003 M_w 6.8 Chengkung and the 1978 M 7.0 Izu-Oshima-Kinkai earthquakes? The recurrences of radon anomalies precursory to local main earthquakes are also essential for radon to be a reliable precursor. Unfortunately for the 1978 radon anomaly, there was no anomalous radon decline with the succeeding earthquake of magnitude 6.7, which occurred on June 29, 1980 (Wakita 1996). These questions and constructive criticism deserve further research to improve earthquake prediction.

Radon anomalies observed before the 2003 M_w 6.8 Chengkung earthquake (Kuo et al. 2006) and before the 1978 Izu-Oshima-Kinkai earthquake of magnitude 7.0 (Wakita et al. 1980) are both encouraging for earthquake prediction research. The above constructive criticism also motivates the researcher to pursue the following goals:

1. in-situ radon-volatilization mechanism to interpret anomalous decrease in groundwater-radon precursory to an earthquake
2. suitable geological conditions to site a well for monitoring anomalous declines in groundwater-radon concentration
3. recurrences of radon anomalies precursory to local main earthquakes
4. confirmation of in-situ radon-volatilization mechanism

The in-situ volatilization mechanism will be presented in Chapter 4 to explain the phenomena of anomalous declines in groundwater radon observed precursory to local main earthquakes. The recurrences of radon anomalies precursory to local main earthquakes are essential for the consistency of earthquake prediction. Very few information about recurrent radon anomalies is reported in the literature. In Chapter 5, recurrences of radon anomalies observed at Antung D1 well will be presented. Well pumping tests in Chapter 5 also help confirm that radon anomalies observed at Antung well D1 are induced by in-situ radon volatilization and not induced by groundwater mixing.

The 2003 M_w 6.8 Chengkung earthquake occurred on the Chihshang Fault which has a faulting surface extending about 30 km in depth and dips approximately 50° to southeast. The Antung hot spring is about 3 km southeast of the Chihshang Fault (or, the Longitudinal Valley Fault). Based on the geology of Antung hot spring in Chapter 4, suitable geological conditions will also be proposed to site a well for monitoring anomalous declines in groundwater-radon concentration. The proposed geological conditions will be tested by monitoring groundwater radon at another site, Paihe Spring, in Chapter 6.

References

Angelier J, Chu HT, Lee JC (1997) Shear concentration in a collision zone: kinematics of the active Chihshang Fault, Longitudinal Valley, eastern Taiwan. Tectonophysics 274:117–143

Angelier J et al (2000) Active faulting and earthquake hazard: the case study of the Chihshang fault, Taiwan. J Geodyn 29:151–185

Hsu TL (1962) Recent faulting in the Longitudinal Valley of eastern Taiwan. Mem Geol Soc China 1:95–102

Kuo T et al (2006) Anomalous decrease in groundwater radon before the Taiwan M6.8 Chengkung earthquake. J Environ Radioact 88:101–106

Lee JC et al (2005) Monitoring active fault creep as a tool in seismic hazard mitigation: insights from creepmeter study at Chihshang, Taiwan. CR Geosci 337:1200–1207

Noguchi M (1964) New method of radon activity measurement with liquid scintillator. Radioisotopes 13(5):362–367

Wakita H et al (1980) Radon anomaly: A possible precursor of the 1978 Izu-Oshima-kinkai earthquake. Science 207(4433):882–883

Wakita H (1996) Geochemical challenge to earthquake prediction. Proc Natl Acad Sci USA 93:3781–3786

Yu SB, Kuo LC (2001) Present-day crustal motion along the Longitudinal Valley Fault, eastern Taiwan. Tectonophysics 333(1–2):199–217

Chapter 4
A Physical Mechanism of Groundwater-Radon Volatilization

Abstract In Chap. 3, we presented the anomalous radon decline observed at Antung hot spring before the 2003 M_w 6.8 Chengkung Earthquake. This chapter presents a physical mechanism, in-situ volatilization mechanism, to interpret anomalous decrease in groundwater-radon precursory to an earthquake. The volatilization mechanism extends naturally from dissolved radon to other dissolved gas components. With the help of the geology at Antung spring, this chapter also outlines suitable geological conditions to site groundwater-radon monitoring wells for catching radon anomalies.

Keywords Radon anomaly · Groundwater · Earthquake · In-situ volatilization mechanism

4.1 Introduction

The radon concentration in groundwater is proportional to the uranium content in adjacent rocks in an aquifer. Because of radon's short recoil length ($3 * 10^{-8}$ m), only atoms produced at the surface of rock grains are released to the surrounding groundwater. Therefore, the radon concentration in groundwater is dependent on both the surface area and the uranium content of aquifer rocks (Torgersen et al. 1990). Before the occurrence of an earthquake, when regional stress increases, the formation of micro-cracks in rock masses could cause an increase in the surface area of rocks. Therefore, radon concentration often rises in the groundwater. For example, Igarashi et al. (1995) reported anomalous increases in the concentration of groundwater-radon precursory to the 1995 Kobe earthquake in Japan. Most radon (Rn-222) anomalies before earthquake occurrence also show increases in radon-concentration precursory to a rupture (Hauksson 1981). Few anomalies manifested decreases in radon before earthquake occurrence. Geological conditions and physical mechanisms are seldom available in the literature for interpreting anomalous decreases in radon prior to earthquakes. Radon-222 is a radioactive nuclide with a half-life of 3.8 days. The transport behavior of radon in geological environments is governed by physical processes such

M. T. Kuo, *Groundwater Radon in the Taiwan Subduction Zone*, Advances in Geological Science,
https://doi.org/10.1007/978-981-99-5350-9_4

as fluid advection, diffusion, partition between liquid and gas phases, and radioactive decay. Radon is a chemically inert gas. Groundwater-dissolved radon tends to escape from groundwater to the gas bubbles in aquifer. The partition of radon between water and gas phases in aquifer can be quantified by Henry's law. Radon partition of radon between water and gas phases in aquifer can cause anomalous declines in the concentration of groundwater-radon precursory to large local earthquakes. The purpose of this chapter is to investigate a physical mechanism, or, in-situ radon-volatilization mechanism (Kuo et al. 2006) for interpreting the phenomena of anomalous decrease in groundwater radon observed prior to the 2003 Chengkung earthquake.

The Antung hot spring is situated in a small fractured aquifer of tuffaceous sandstone surrounded by ductile mudstone. Given these geological conditions, we hypothesize that gas saturation develops in newly created cracks preceding the earthquake. Based on radon partition between liquid and gas phases, groundwater-dissolved radon volatilizes from groundwater into the gas phase, which may explain the anomalous decrease of radon precursory to the 2003 Chengkung earthquake (Kuo et al. 2006). The in-situ radon-volatilization mechanism is also useful to explain other radon anomalies observed in the world such as the well-documented anomalous decrease in groundwater-radon concentration before the 1978 Izu-Oshima-Kinkai earthquake of magnitude 7.0 (Wakita et al. 1980).

In a low-porosity fractured aquifer with limited recharge (or, under undrained conditions) gas bubbles develop in rock fractures prior to local main earthquakes. Like groundwater-dissolved radon, other groundwater-dissolved gas components, such as methane, ethane, etc., also volatilize from groundwater into gas bubbles. Therefore, the in-situ volatilization mechanism is applicable to groundwater-dissolved radon and other groundwater-dissolved gas components.

4.2 Suitable Geological Conditions

The 2003 M_w 6.8 Chengkung earthquake, which occurred on December 10, 2003, was the strongest earthquake near the Chengkung area in eastern Taiwan since 1951. The Antung radon-monitoring well D1 was located 24 km from the epicenter. Approximately 65 days prior to the 2003 M_w 6.8 Chengkung earthquake, groundwater radon started to decrease from a background level of 787 pCi/L to a minimum concentration of 326 pCi/L. Understanding the geology of well D1 at the Antung hot spring is helpful to find out suitable geological conditions to site a well for monitoring anomalous declines in groundwater-radon concentration.

Figure 4.1 shows the geological map and cross section near the radon-monitoring well D1, which is about 3 km southeast of the Longitudinal Valley Fault (Chen and Wang 1996). The Longitudinal Valley Fault is the present-day plate suture between the Eurasian and the Philippine Sea plates. The Longitudinal Valley Fault (Hsu 1962) ruptured during two 1951 earthquakes of magnitude (M) 6.2 and (M) 7.0. Four stratigraphic units are present near the Antung hot spring. The Tuluanshan Formation consists of Miocene volcanic rocks such as lava and volcanic breccia as well

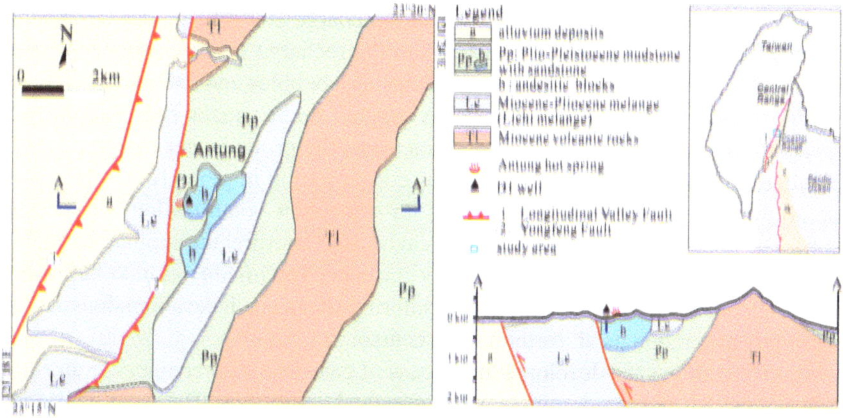

Fig. 4.1 Geological map and cross section near well D1 in Antung hot spring (From Kuo 2022 with permission of *Groundwater*, National Ground Water Association) (B: tuffaceous andesitic blocks; filled black triangle: well D1; ①: Chihshang, or, Longitudinal Valley Fault, ②: Yongfeng Fault)

as tuffaceous sandstone. The Fanshuliao and Paliwan Formations consist of Plio-Pleistocene mudstone turbidites with rhythmic sandstone. The Lichi mélange occurs as a highly deformed mudstone that is characterized by penetrative foliation visible in outcrop.

The Antung hot spring is situated in an exotic brittle tuffaceous sandstone block surrounded by a ductile mudstone of the Paliwan Formation and the Lichi mélange (Chen and Wang 1996). The exotic block extends about 1 or 2 km. The aquifer size at Antung well D1 is small. Minor faults and joints are well developed with an average thickness of 40 m in the tuffaceous sandstone block displaying intensively brittle deformation. It is possible that these fractures reflect deformation by the nearby faults. Groundwater flows through the fault zone and is then into the block along the minor fractures. Aquifer recharge at Antung well D1 is weak, which can be approximated by undrained conditions. The unique geological conditions near Antung well D1 were instrumental in catching the radon anomaly precursory to the 2003 M_w 6.8 Chengkung earthquake (Kuo et al. 2006). A small low-porosity brittle aquifer surrounded by a ductile formation in undrained conditions, such as Antung well D1, is a suitable site to detect radon anomalies consistently.

4.3 In-Situ Radon-Volatilization Mechanism

Figure 4.1 shows that the Antung hot spring is situated at the hanging wall along the Yongfeng fault and the Longitudinal Valley Fault. Both faults are thrust faults. Along a thrust fault, rock dilation is likely to take place at the hanging wall (Doglionia et al. 2011). When the regional stress increases to about half the fracture stress, dilation of brittle rock initiates and micro-cracks develop in aquifer rock (Brace et al. 1966;

Nur 1972; Scholz et al. 1973). When aquifer recharge is weak, the development of new cracks could occur at a rate faster than the recharge of pore water in a small brittle aquifer. Prior to local large earthquakes, gas bubbles and two phases (gas and water) are likely to develop in the newly created micro-cracks near Antung well D1. As the gas phase develops in rock fractures, the groundwater-dissolved radon volatilizes into the gas phase and the radon concentration in groundwater decreases. The above mechanism is referred to as "in-situ radon volatilization" (Kuo et al. 2006). The mechanism provides the physical basis to interpret the anomalous decline in groundwater radon before earthquake occurrence. The in-situ radon-volatilization mechanism also helps select a suitable monitoring site to catch the anomalous decline in groundwater radon as an earthquake precursor.

Figure 4.2 shows the development process of gas-bubbles precursory to an earthquake in a small fractured aquifer with undrained conditions. The small undrained fractured aquifer situates in a brittle rock, which is surrounded by a ductile formation. Before any precursory phenomenon appears (Stage 1), Fig. 4.2a shows that there is only a water phase in the aquifer. When the regional tectonic stress continues to increase and aquifer recharge is weak, Fig. 4.2b shows that dilation of brittle rock could occur at a faster rate than the rate of groundwater recharging into the newly created micro-cracks. As a result, gas bubbles and two phases (gas and water) develop in the aquifer. The radon in groundwater volatilizes into the gas bubbles and the radon concentration in groundwater decreases. Before any precursory phenomenon appears, Fig. 4.2a shows that the aquifer is water-saturated and there is only water phase present in the fracture space. During the period of radon anomaly (Stages 2 and 3), there are two phases (gas and water) in the aquifer. Figure 4.2b shows that gas bubbles are present in the fracture space as a distinct gas phase. In Stage 2, the dilation rate of brittle rock is faster than the recharge rate of groundwater, and vice versa in Stage 3. Mass transfer of radon occurs between water and gas phases during Stages 2 and 3. At the end of radon anomaly, the aquifer becomes water-saturated again and the groundwater-radon concentration recovers to the previous background level before the earthquake.

Figure 4.2c shows the radon anomaly observed at Antung well D1 prior to the 2003 Chengkung earthquake. The radon anomaly clearly progresses in a sequence of three stages. Stage 1 is buildup of elastic strain. During Stage 1, the radon concentration in groundwater was fairly stable at 787 ± 42 pCi/L. When the regional stress continued to increase, dilation of brittle rock masses occurred. Stage 2 is the development of cracks and gas bubbles in the brittle aquifer. During Stage 2, the radon in groundwater volatilizes into the gas bubbles. The radon concentration of groundwater starts to decrease and reaches a minimum value at 326 ± 9 pCi/L. Stage 3 is an influx of groundwater. Stage 3 starts at the point of minimum radon concentration. During Stage 3, the radon concentration in groundwater increases and recovers to the previous background level before the earthquake.

Radon volatilization from groundwater into the gas phase can explain the anomalous decreases of radon precursory to the earthquakes (Kuo et al. 2006). For a confined aquifer with undrained conditions, the in-situ radon-volatilization model, Eq. (4.1), was developed as follows (Kuo et al. 2006).

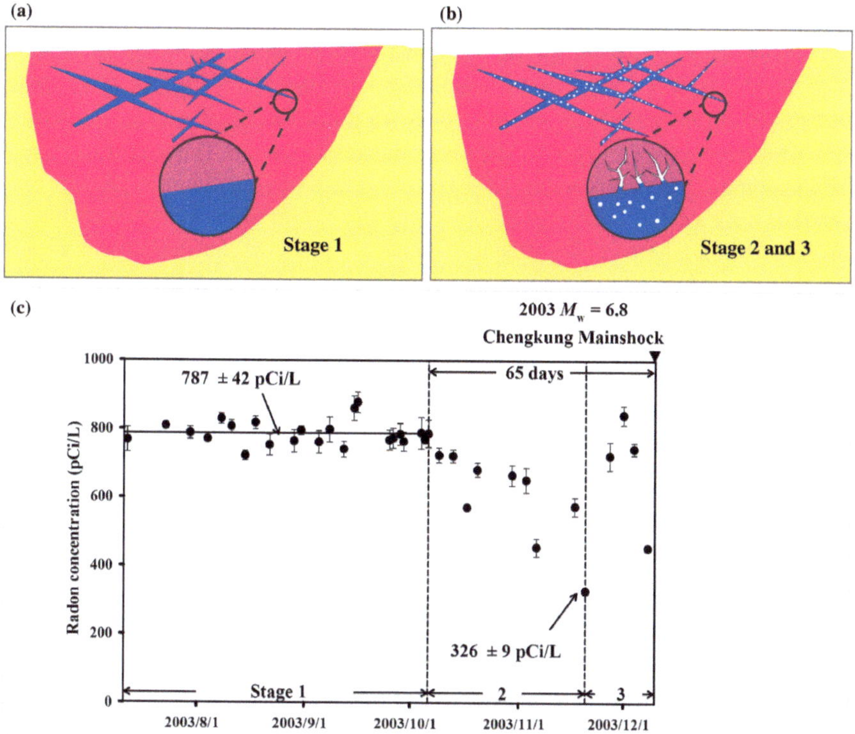

Fig. 4.2 Development of gas bubbles in a small fractured aquifer with undrained conditions. **a** Stage 1, the aquifer is water-saturated (groundwater painted in blue; brittle rock in pink; ductile formation in yellow). **b** Stages 2 and 3, micro-cracks and gas bubbles develop (gas bubbles: white circles; micro-cracks: white branch; dilation of brittle rock shown in enlarged circle). **c** Radon-concentration data at Antung well D1 prior to 2003 Chengkung earthquake (From Kuo 2022 with permission of *Groundwater*)

$$C_0 = C_w(H \times S_g + 1) \tag{4.1}$$

where C_0 is the initial radon concentration in groundwater precursory to each radon anomaly, pCi/L; C_w is the radon minimum in groundwater observed in well D1 during an anomalous decline, pCi/L; S_g is gas saturation, fraction; H is Henry's coefficient for radon at aquifer temperature, dimensionless. Equation (4.1) correlates the observed decline in groundwater radon with the gas saturation developed in a confined aquifer. Based on the Weigel equation (1978), the Henry's coefficients (H) at aquifer temperature (60 °C) is 7.91 for radon. Using the data of radon background and minima from Fig. 4.2c and Eq. (4.1), the gas saturation developed in Antung fractured aquifer can be determined, $S_g = 17.9\%$ prior to the 2003 M_w 6.8 Chengkung earthquake.

Variations of radon concentration (radon-222) in groundwater have been applied as possible precursor in earthquake prediction studies (Noguchi and Wakita 1977;

Wakita et al. 1980; Shapiro et al. 1980; Hauksson and Goddard 1981; Igarashi et al. 1995; Wakita 1996; Roeloffs 1999; Trique et al. 1999, Kuo et al. 2006; Erees et al. 2007; Kuo et al. 2013; Tsunomori and Tanaka 2014; Tarakçı et al. 2014; Morales-Simfors et al. 2020). In this book, we focus on the anomalous declines in groundwater-radon precursory to large earthquakes in Taiwan. The well-documented radon anomaly precursory to the 1978 Izu-Oshima-Kinkai earthquake (Wakita et al. 1980) is lack of the reproducibility to measure recurrences of radon anomalies (Wakita 1996). There is also a lack of plausible physical mechanisms to explain the observed decline in groundwater radon (Wakita 1996). In Chap. 5, we present recurrences of radon anomalies recorded at Antung well D1 precursory to main local thrust-type earthquakes from 2003 to 2010. To corroborate the validity of suitable geological conditions, groundwater radon will be monitored at another small low-porosity fractured aquifer, Paihe Spring, in Chap. 6. The mechanism of in-situ radon volatilization will be applied to explain the anomalous declines in groundwater-radon precursory to the 1978 Izu-Oshima-Kinkai earthquake in Chap. 7.

4.4 In-Situ Volatilization Mechanism for Groundwater-Dissolved Gas

In a low-porosity fractured aquifer with limited recharge (or, under undrained conditions) gas bubbles develop in rock fractures prior to local main earthquakes. Groundwater dissolved radon then volatilizes from groundwater into gas bubbles. Likewise, other groundwater-dissolved gases, such as methane, ethane, etc., also volatilize from groundwater into gas bubbles. Simultaneous concentration declines in groundwater-dissolved radon and methane were observed precursory to the 2008 M_w 5.4 Antung earthquake (Kuo et al. 2010). Again, simultaneous concentration declines in groundwater-dissolved radon, methane, and ethane were observed precursory to the 2011 M_w 5.0 Chimei earthquake (Kuo et al. 2013).

For a low-porosity confined aquifer, the in-situ radon-volatilization model, Eq. (4.1), can be modified for groundwater-dissolved methane and ethane as follows (Kuo et al. 2013).

$$C_{0,\,Me} = C_{w,\,Me}(H_{Me} \times S_g + 1) \tag{4.2}$$

$$C_{0,\,Et} = C_{w,\,Et}(H_{Et} \times S_g + 1) \tag{4.3}$$

where $C_{0,\,Me}$ is the observed background methane concentration in groundwater when the gas saturation remains at zero, mg/L; $C_{w,\,Me}$ is the observed methane concentration remaining in groundwater during the process of rock dilatancy, mg/L; S_g is gas saturation, %; H_{Me} is Henry's coefficient for methane, dimensionless; $C_{0,\,Et}$ is the observed background ethane concentration in groundwater when the gas saturation remains at zero, mg/L; $C_{w,\,Et}$ is the observed ethane concentration remaining in groundwater during the process of rock dilatancy, mg/L; S_g is gas saturation, %; H_{Et} is Henry's coefficient for ethane, dimensionless. Equations (4.1), (4.2), and (4.3) correlate the observed decline in groundwater-dissolved radon, methane, and ethane with the gas saturation developed in a confined aquifer, respectively. The concentrations of groundwater-dissolved radon, methane, and ethane are affected by two key parameters, Henry's constant and the gas saturation developed in a low-porosity fractured aquifer precursory to a local main earthquake.

References

Brace WF, Paulding BW Jr, Scholz CH (1966) Dilatancy in the fracture of crystalline rocks. J Geophys Res 71(16):3939–3953

Chen WS, Wang Y (1996) Geology of the coastal range, eastern Taiwan. Geology of Taiwan 7

Doglionia C et al (2011) Role of the brittle–ductile transition on fault activation. Phys Earth Planet Inter 184:160–171

Erees FS et al (2007) Radon concentrations in thermal waters related to seismic events along faults in the Denizli Basin, Western Turkey. Radiat Meas 42:80–86

Hauksson E (1981) Radon content of groundwater as an earthquake precursor: evaluation of worldwide data and physical basis. J Geophys Res 86(B10):9397–9410

Hauksson E, Goddard JG (1981) Radon earthquake precursor studies in Iceland. J Geophys Res: Solid Earth 86(B8):7037–7054

Hsu TL (1962) Recent faulting in the Longitudinal Valley of eastern Taiwan. Mem Geol Soc China 1:95–102

Igarashi G et al (1995) Ground-water radon anomaly before the Kobe earthquake in Japan. Science 269:60–61

Kuo MCT (2022) Study of water relative permeability in fractures using well tests and radon: gas bubbles effect. Ground Water 60(4):510–517

Kuo MCT et al (2006) A mechanism for anomalous decline in radon precursory to an earthquake. Ground Water 44(5):642–647

Kuo T et al (2010) Simultaneous declines in radon and methane precursory to 2008 MW 5.0 Antung earthquake: corroboration of in-situ volatilization. Nat Hazard 54(2):367–372

Kuo T et al (2013) Concurrent concentration declines in groundwater-dissolved radon, methane and ethane precursory to 2011 MW 5.0 Chimei earthquake. Radiat Meas 58:121–127

Morales-Simfors N, Wyss RA, Bundschuh J (2020) Recent progress in radon-based monitoring as seismic and volcanic precursor: a critical review. Crit Rev Environ Sci Technol 50(10):979–1012

Noguchi M, Wakita H (1977) A method for continuous measurement of radon in groundwater for earthquake prediction. J Geophys Res 82(8):1353–1357

Nur A (1972) Dilatancy, pore fluids, and premonitory variations of ts/tp traval times. Bull Seismol Soc Am 62(5):1217–1222

Roeloffs E (1999) Radon and rock deformation. Nature 399:104–105

Scholz CH, Sykes LR, Aggarwal YP (1973) Earthquake prediction: a physical basis. Science 181(4102):803–810

Shapiro MH, Melvin JD, Tombrello TA (1980) Automated radon monitoring at a hard-rock site in the southern California transverse ranges. J Geophys Res 85:3058–3064

Tarakçı M et al (2014) Investigation of the relationships between seismic activities and radon level in western Turkey. Appl Radiat Isot 83:12–17

Torgersen T, Benoit J, Mackie D (1990) Controls on groundwater Rn-222 concentrations in fractured rock. Geophys Res Lett 17(6):845–848

Trique M et al (1999) Radon emanation and electric potential variations associated with transient deformation near reservoir lakes. Nature 399:137–141
Tsunomori F, Tanaka H (2014) Anomalous change of groundwater radon concentration monitored at Nakaizu well in 2011. Radiat Meas 60:35–41
Wakita H (1996) Geochemical challenge to earthquake prediction. Proc Natl Acad Sci USA 93:3781–3786
Wakita H et al (1980) Radon anomaly: a possible precursor of the 1978 Izu-Oshima-kinkai earthquake. Science 207(4433):882–883
Weigel F (1978) Radon. Chemiker-Zeitung 102:287–299

Chapter 5
Recurrences of Radon Anomalies Precursory to Local Main Earthquakes at Antung Hot Spring

Abstract Recurrences of radon anomalies are essential in earthquake prediction. In this chapter, we will present recurrent anomalous declines in groundwater-radon concentration consistently recorded at Antung prior to local main earthquakes that occurred on the Longitudinal Valley Fault between 2003 and 2010. The formation of gas bubbles is necessary for the process of in-situ radon volatilization to occur. Well tests can provide field evidence regarding the in-situ development of gas bubbles in water-saturated brittle rock prior to the 2008 M_w 5.4 Antung earthquake. In addition, aquifer transmissivity determined from well tests can complement groundwater radon as an earthquake precursor.

Keywords Precursors · Groundwater radon · Aquifer transmissivity · Gas bubbles · The Longitudinal Valley Fault

5.1 Introduction

Recurrences of radon anomalies are essential in earthquake prediction. Between 2003 and 2010, anomalous declines in the concentration of groundwater radon were consistently recorded at Antung well D1 precursory to four main earthquakes near the Longitudinal Valley Fault (the 2003 M_w 6.8 Chengkung, 2006 M_w 6.1 and M_w 5.9 Taitung, and 2008 M_w 5.4 Antung earthquakes). A small low-porosity fractured aquifer can be an effective natural strainmeter for earthquake warnings. When regional stress increases, micro-cracks in brittle rock could form at a faster rate than the rate of groundwater recharge in the aquifer with undrained conditions. As a result, gas bubbles develop and induce detectable earthquake precursors. A suitable geological site cited here is Antung andesite spring in the southern Taiwan subduction zone.

The presence of gas bubbles in a fractured aquifer is required for the volatilization of groundwater-dissolved radon. Well tests provided indirect field evidence regarding the development of gas bubbles in water-saturated fractured rock prior to the 2008 M_w 5.4 Antung earthquake. Anomalous declines in both groundwater radon and

M. T. Kuo, *Groundwater Radon in the Taiwan Subduction Zone*, Advances in Geological Science,
https://doi.org/10.1007/978-981-99-5350-9_5

aquifer transmissivity were observed prior to the 2008 M_w 5.4 Antung earthquake. The findings are remarkable because the change in radon concentration coincided qualitatively and quantitatively with a similar change in aquifer transmissivity (Kuo 2022). Both groundwater radon and aquifer transmissivity can be monitored together at the same well as earthquake precursors globally in the subduction zone.

Anomalous declines in groundwater radon, greater than 30%, were consistently recorded at Antung andesite spring precursory to four large main earthquakes in southeastern Taiwan (the 2003 M_w 6.8 Chengkung, 2006 M_w 6.1 and M_w 5.9 Taitung, and 2008 M_w 5.4 Antung earthquakes). Table 5.1 summarizes radon observations precursory to the four shallow main earthquakes which occurred near Antung for the period between 2003 and 2010.

Well tests were also conducted to estimate aquifer transmissivity before and after the 2008 Antung earthquake. Anomalous declines in both groundwater-radon concentration and aquifer transmissivity were observed precursory to the 2008 Antung earthquake. It is fortunate to discover that aquifer transmissivity can complement groundwater radon as an earthquake precursor. The precursory strain in the crust shows a decrease in aquifer transmissivity instead of an increase. A physical basis is presented in this chapter to explain the anomalous decline in aquifer transmissivity recorded precursory to the 2008 Antung earthquake. Based on the observed declines in aquifer transmissivity and groundwater radon, mathematical models are utilized to quantify the gas saturation required to develop in rock fractures prior to the 2008 Antung earthquake respectively.

The first objective of this chapter is to present recurrent anomalous declines in groundwater radon consistently recorded at Antung prior to local main earthquakes that occurred near Antung hot springs for the period between 2003 and 2010 (M_w range 5.4–6.8). As shown in Table 5.1, four main large shallow earthquakes occurred during these 7 years with epicenters located within a distance of 52 km from Antung.

Table 5.1 Important parameters of the radon anomalies precursory to four main earthquakes recorded at Antung well D1 near the Longitudinal Valley Fault in the southern Taiwan subduction zone

Event	Main earthquake	Magnitude (M_W)	Radon background (pCi/L)	Radon minimum (pCi/L)	Percent of decline (%)	Precursor time (day)	Epicenter distance (km)
1	2003 Chengkung	6.8	787	326	59	65	24
2	2006 Taitung (April 1)	6.1	762	371	51	61	52
3	2006 Taitung (April 15)	5.9	762	371	51	75	47
4	2008 Antung	5.4	700	480	31	56	13

The radon anomalies consistently observed at Antung substantiate that dilation of brittle rock can be a repeatable process for many cycles of compressive stress (Brace et al. 1966; Nur 1972; Scholz et al. 1973; Zoback and Byerlee 1975). With the help of Antung case study, we demonstrate the reproducibility to measure radon anomalies precursory to local main earthquakes at a suitable geological site.

The second objective of this chapter is to provide field evidence regarding the development of gas bubbles precursory to the 2008 Antung earthquake. The development of gas bubbles in water-saturated brittle rock prior to an earthquake is the basic hypothesis of the in-situ radon-volatilization mechanism (Kuo et al. 2006). Direct observation of gas bubbles in aquifers under field conditions is very difficult. Instead, by using well-pumping tests, this chapter provides indirect evidence regarding the development of gas bubbles in a fractured aquifer precursory to an earthquake. Aquifer transmissivity can be monitored by well-pumping tests. The presence of gas bubbles in fractures reduces the water's relative permeability and aquifer transmissivity. Therefore, by detecting anomalous declines in aquifer transmissivity, we can infer the development of gas bubbles as a distinct gas phase in fractures.

The main goal of this chapter is to decipher the precursory process of gas-bubble development in a small fractured aquifer with undrained conditions. Both declines in groundwater radon and aquifer transmissivity are two precursory phenomena having a common effect of gas bubbles. Analytical models are used to quantify the effect of gas bubbles on the precursory declines in groundwater radon and aquifer transmissivity.

5.2 Recurrences of Radon Anomalies Observed at Antung D1 Well

Monitoring of groundwater radon began in July 2003 at well D1 located at the Antung hot spring. As shown in Fig. 5.1, the Antung well D1 is located about 3 km southeast of the Longitudinal Valley Fault, which is part of the present-day plate suture between Eurasia and Philippine Sea plates. It ruptured during two 1951 earthquakes of magnitudes M 6.2 and M 7.0 (Hsu 1962). There are four local main earthquakes that occurred near Antung well D1 for the period between July 2003 and September 2010, as also shown in Fig. 5.1. Anomalous decreases in the concentration of groundwater radon were recurrently and consistently observed prior to these four main earthquakes: the 2003 M_w 6.8 Chengkung, 2006 M_w 6.1 and M_w 5.9 Taitung, and 2008 M_w 5.4 Antung earthquakes. The 2006 M_w 5.9 Taitung earthquake that occurred on April 15, 2006, was triggered by the 2006 M_w 6.1 Taitung earthquake that occurred on April 1, 2006 (Wu et al. 2006).

Table 5.1 summarizes the observed radon background, radon minimum, radon-decline percent, precursor time, and epicenter distance from Antung well D1 for radon anomalies precursory to large main earthquakes in southeastern Taiwan for

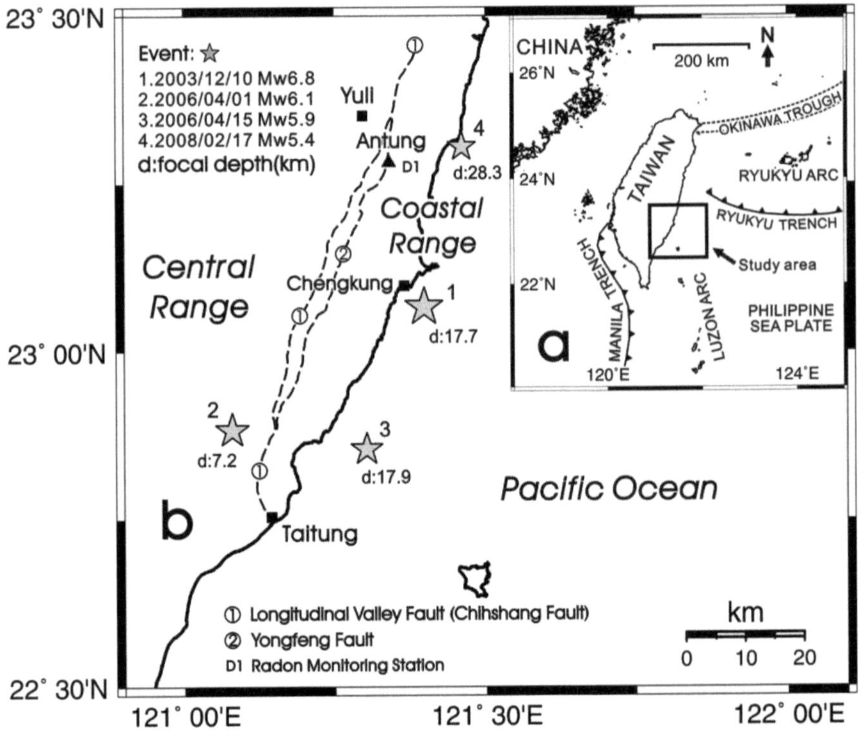

Fig. 5.1 Map of the epicenters of the mainshocks that occurred on December 10, 2003, April 1 and 15, 2006, and February 17, 2008, near the Antung hot spring. **a** The geographical location of Taiwan. **b** Study area near the Antung hot spring (filled stars: mainshocks, filled triangle: Antung well D1) (From Kuo et al. 2011)

the period between 2003 and 2010. As shown in Table 5.1, radon concentration at Antung D1 well decreased from background levels of 787 ± 42, 762 ± 57, and 700 ± 57 pCi/L to minima of 326 ± 9, 371 ± 9, and 480 ± 43 pCi/L, respectively, precursory to the 2003 M_w 6.8 Chengkung, 2006 M_w 6.1 and M_w 5.9 Taitung, and 2008 M_w 5.4 Antung earthquakes. As the earthquake magnitude increases, the radon minimum decreases. This is an important observation because it implies the potential application of radon precursor to measure crustal strain quantitatively.

Figure 5.2 shows the observed concentration of groundwater radon versus the date from July 2003 to September 2010. Radon-concentration errors are ± 1 standard deviation after simple averaging of triplicates. Recurrent radon anomalies were observed at Antung D1 well precursory to the four local main earthquakes (the 2003 M_w 6.8 Chengkung, 2006 M_w 6.1 and M_w 5.9 Taitung, and 2008 M_w 5.4 Antung earthquakes). The M_w 5.9 Taitung earthquake that occurred on April 15, 2006, was triggered by stress transfer in response to the 2006 M_w 6.1 Taitung earthquake that occurred on April 1, 2006. All three recurrent anomalies observed at Antung D1

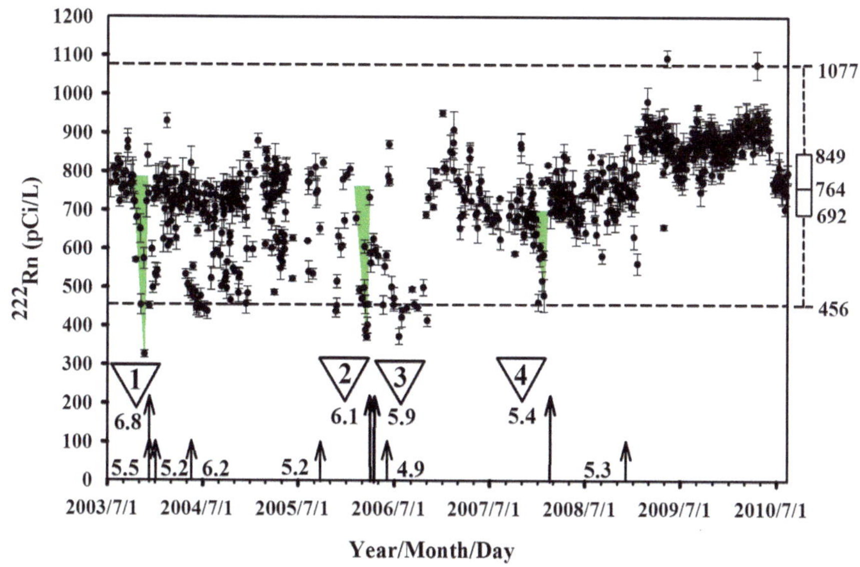

Fig. 5.2 Observed radon concentration at Antung well D1 for the period between July 2003 and September 2010. Numbers in open inverted triangles correspond to earthquake events in Table 5.1 (long arrows: mainshocks; short arrows: aftershocks; earthquake magnitude M_w shown beside arrows; green inverted triangles: v-shaped patterns shown in Fig. 5.3) (From Kuo et al. 2011)

well follow the same v-shaped progression and are marked with green inverted triangles in Fig. 5.2. Environmental records such as atmospheric temperature, barometric pressure, and rainfall were examined to confirm that the radon anomalies were not attributed to these environmental factors.

The box-and-whisker plot, on the right-hand side of Fig. 5.2, shows the median (50th percentile, 764 pCi/L) as a center bar, and the quartiles (25th and 75th percentiles, 692 pCi/L and 849 pCi/L) as a box. The upper and lower whiskers (456 pCi/L and 1077 pCi/L) cover all but the most extreme values in the data set. Using the lower whisker, 456 pCi/L is estimated as the threshold concentration of anomalous radon minima at Antung D1 well. For the 2003 M_w 6.8 Chengkung and 2006 M_w 6.1 and M_w 5.9 Taitung earthquakes, the recorded radon anomalous minima are low enough to be clearly distinguished from the background noise. On the other hand, the radon minimum recorded prior to the 2008 M_w 5.4 Antung earthquake is at the threshold and may be masked by the noisy background. The threshold concentration of anomalous radon minima also suggests the sensitivity of using groundwater radon as an earthquake precursor at the Antung D1 well.

Figure 5.2 shows large background variations in radon data between the 2003 M_w 6.8 Chengkung and 2006 M_w 6.1 Taitung earthquakes. Following the 2003 M_w 6.8 Chengkung earthquake, four aftershocks with magnitudes (M_w) of 5.5, 5.2, 6.2, and 5.2 occurred on 12/11/2003, 1/1/2004, 5/19/2004, and 9/26/2005, respectively. Large scatter in radon data between the 2003 M_w 6.8 Chengkung and 2006 M_w 6.1 Taitung earthquakes can be related to these aftershocks. The 2006 M_w 6.1 Taitung earthquake that occurred on April 1, 2006, also triggered the M_w 5.9 Taitung earthquake that occurred on April 15, 2006. Based upon magnitude and location, the earthquake of magnitude M_w 4.9 that occurred on 6/4/2006 can be considered as an aftershock of the 2006 M_w 6.1 Taitung earthquake. The M_w 4.9 aftershock following the 2006 M_w 6.1 Taitung earthquake also caused a large scatter in radon data. The background variation in radon data following the 2008 M_w 5.4 Antung earthquake can also be attributed to aftershocks, such as a local earthquake of magnitude M_w 5.3 that occurred on 12/2/2008. In summary, Fig. 5.2 reveals the larger magnitude of the main shock, the more aftershocks follow and the larger background variations in radon data occur.

The radon anomalies are marked in Fig. 5.2 by green inverted triangles and are expanded in Fig. 5.3. Figure 5.3a, b, and c show that the radon anomalies clearly progress in a sequence of three stages. Stage 1 is a buildup of elastic strain. During Stage 1, the radon concentration in groundwater was fairly stable; there was a slow, steady increase in regional stress. Stage 2 is the development of cracks and gas bubbles in the brittle aquifer. The Antung hot spring is a small fractured aquifer with limited recharge. Under these geological conditions and as the regional stress increased, dilation of brittle rock masses occurred at a rate faster than the rate at which groundwater could recharge into the newly created micro-cracks (Brace et al. 1966; Nur 1972; Scholz et al. 1973). Gas saturation and two phases (gas and liquid) developed in the rock cracks during Stage 2. The radon in groundwater escaped into the gas phase and the radon concentration in groundwater decreased. Stage 3 starts at the point of minimum radon concentration. During Stage 3, the radon concentration in groundwater increases and recovers to the previous background level before the main earthquake.

Undrained conditions are reasonable approximations in Stage 2. Under undrained conditions, the radon concentration in groundwater decreases from a background level to a minimum concentration. Figure 5.3a, b, and c show that Stage 2 prior to the 2003 M_w 6.8 Chengkung, 2006 M_w 6.1 and M_w 5.9 Taitung, and 2008 M_w 5.4 Antung earthquakes are 45, 47, and 31 days, respectively.

For a confined fractured aquifer with undrained conditions, in-situ radon-volatilization model, Eq. (5.1), is developed as follows to correlate the observed decline in groundwater radon with the gas saturation developed in the aquifer (Kuo et al. 2006).

$$C_0 = C_w(H \times S_g + 1) \tag{5.1}$$

where C_0 is the initial radon concentration in groundwater precursory to each radon anomaly, pCi/L; C_w is the radon minimum in groundwater observed at well D1 during an anomalous decline, pCi/L; S_g is gas saturation, fraction; H is Henry's coefficient for radon at aquifer temperature, dimensionless. The Henry's coefficient (H) at aquifer temperature (60 °C) is 7.91 for radon (Weigel 1978). Using Eq. (5.1) and the data of radon background (C_0) and minima (C_w) from Table 5.1, the gas saturation developed in Antung well D1 aquifer can be estimated, $S_g = 17.9\%, 13.3\%$,

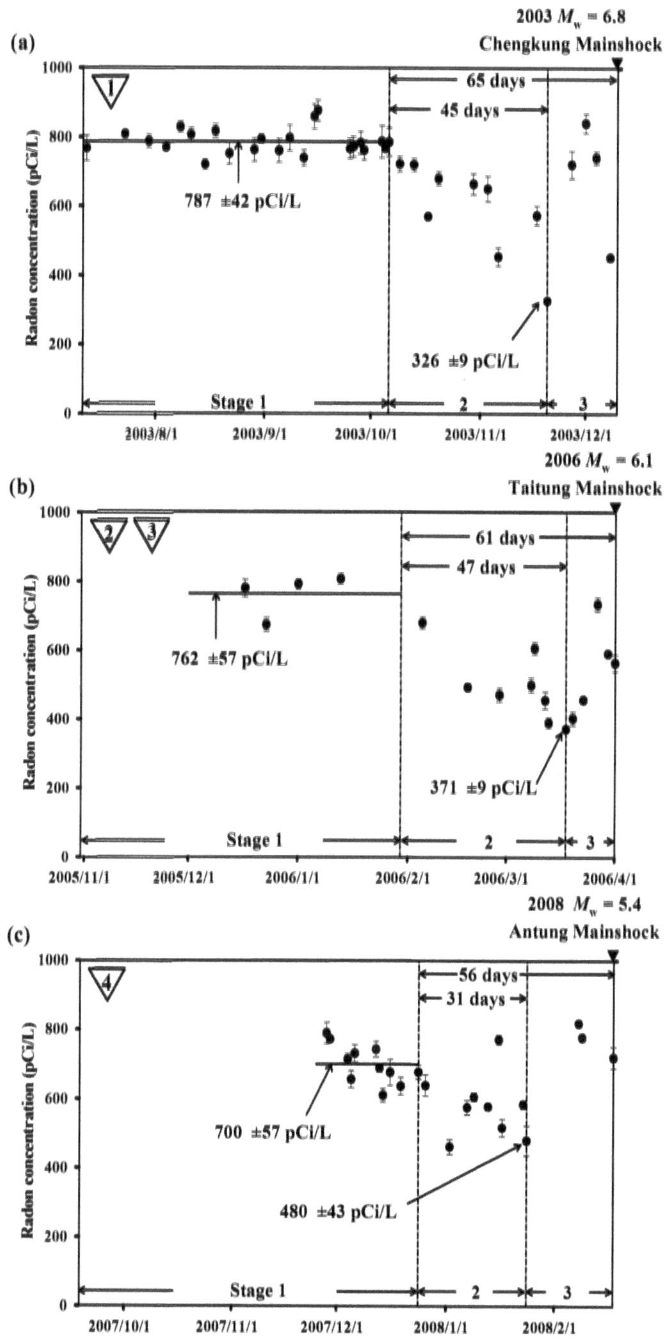

Fig. 5.3 Observed radon anomalies at well D1 prior to. **a** 2003 Chengkung. **b** 2006 April 1 and April 15 Taitung, and **c** 2008 Antung earthquakes. Stages 1, 2, and 3 are defined in the text. Numbers in inverted triangles correspond to earthquake events in Table 5.1

5.8%, prior to the 2003 M_w 6.8 Chengkung, 2006 M_w 6.1 and M_w 5.9 Taitung, and 2008 M_w 5.4 Antung earthquakes, respectively.

The precursor time for radon is defined as the time interval between the moment when the trend of radon concentration starts to decline and the time of occurrence of the earthquake. The precursor time for radon is an important quantitative parameter to make a short-term earthquake warning. Table 5.1 summarizes that the precursor time observed at Antung well D1 is approximately 56–75 days (about 2–3 months) for the main earthquakes occurring on the Longitudinal Valley Fault. Via a basic observation of groundwater radon in a small low-porosity fractured aquifer, prediction of main earthquakes to occur on the Longitudinal Valley Fault is achievable a couple of months in advance.

Variations of radon concentration in groundwater have been applied as a possible precursor in earthquake prediction studies (Kuo et al. 2006; Wakita et al. 1980; Wakita 1996; Shapiro et al. 1980; Kuo et al. 2013; Kuo 2014; Noguchi and Wakita 1977; Hauksson and Goddard 1981; Hauksson 1981; Roeloffs 1999; Erees et al. 2007; Tarakçı et al. 2014; Kuo et al. 2017; Kuo et al. 2018; Morales-Simfors et al. 2020). According to a worldwide survey (Hauksson 1981), more than 80% of radon anomalies associated with earthquakes show increases in radon concentration while only a few anomalies manifested decreases in radon. We focus on the anomalous declines in groundwater radon precursory to main earthquakes in Taiwan (Kuo et al. 2006, 2013), Japan (Wakita et al. 1980; Wakita 1996), and the United States (Shapiro et al. 1980). The well-documented radon-decline anomaly precursory to the 1978 Izu-Oshima-Kinkai earthquake (Wakita et al. 1980; Wakita 1996) is thus far irreproducible. Fortunately, this chapter provides reproducible evidence by recording the recurrences of radon anomalies at Antung precursory to four main earthquakes for the period between 2003 and 2010.

In-situ radon-volatilization mechanism is applied to explain the anomalous declines in groundwater radon precursory to the above main earthquakes (Kuo et al. 2006). The observed anomalous declines in groundwater radon are attributed to the radon volatilization from groundwater into gas bubbles. In-situ radon-volatilization mechanism hypothesizes the development of gas bubbles in water-saturated fractured rock prior to an earthquake (Kuo et al. 2006, 2013). To confirm the mechanism of in-situ radon volatilization, we search for field evidence regarding the development of gas bubbles in the fractured aquifer prior to the earthquake.

5.3 Aquifer Transmissivity as a Complementary Earthquake Precursor*

* Kuo 2022 with permission of *Groundwater*, National Ground Water Association.

The objective of this section is to confirm the mechanism of in-situ radon-volatilization by providing field evidence regarding the development of gas bubbles precursory to the 2008 Antung earthquake. Direct field observation of gas bubbles in aquifers is difficult. Instead, an indirect approach is possible regarding the development of gas bubbles in water-saturated fractured rock precursory to an earthquake. The presence of gas bubbles in the fracture space as a distinct gas phase will interfere with water flow in fractured aquifers. Therefore, the development of gas bubbles in water-saturated fractured rock will reduce the water's relative permeability or aquifer transmissivity (Chen et al. 2004; Persoff and Pruess 1995). By detecting anomalous declines in aquifer transmissivity, we can infer the development of gas bubbles as a distinct gas phase in the fracture space. Aquifer transmissivity can be monitored by well-pumping tests.

A total of 139 pumping tests were conducted at Antung well D1 to monitor the change of aquifer transmissivity, between May 1, 2007, and August 19, 2008. A submersible pump was used in well D1 at a fairly constant flow rate of 205 L/min.

The production interval of well D1 ranges from 167 to 187 m below the ground surface. During the 139 pumping tests, the drawdown was recorded at observation well B which is 46 m from pumping well D1.

The Cooper and Jacob (1946) solution, which is a large-time approximate solution of Theis' (1935) solution, was used to analyze drawdown data recorded at observation well B. Figure 5.4 shows an example of the drawdown data and Jacob semi-log straight line for the pumping test conducted on May 15, 2007. For large values of pumping time (when $t > \frac{25r^2S}{T}$),

$$s = \frac{2.30Q}{4\pi T}\log t - \frac{2.30Q}{4\pi T}\log\frac{r^2s}{2.25T} \tag{5.2}$$

where

s = drawdown, in meters, measured in an observation well due to the constant discharge of a pumping well.

Q = discharge of pumping well, in m^3/min.

T = transmissivity, in m^2/min.

r = distance, in meters, from the pumping well to the observation well.

S = coefficient of storage, dimensionless.

t = time in minutes since pumping started.

Figure 5.4 also shows the Jacob semi-log straight with a slope of 1.0156 m per log cycle. From Eq. (5.2),

$$\frac{2.30Q}{4\pi T} = 1.0156 \text{ m}.$$

Given $Q = 0.205$ m^3/min for the pumping test, the aquifer transmissivity on May 15, 2007, is 0.0369 m^2/min.

Figure 5.5a and b show concurrent anomalous declines in groundwater radon and aquifer transmissivity precursory to the 2008 M_w 5.4 Antung earthquake, respectively. It is significant that the patterns and occurrence times of the radon anomalies closely resemble those of aquifer transmissivity. The similarity implies that these changes were caused by a single tectonic force nearby Antung well D1 prior to the

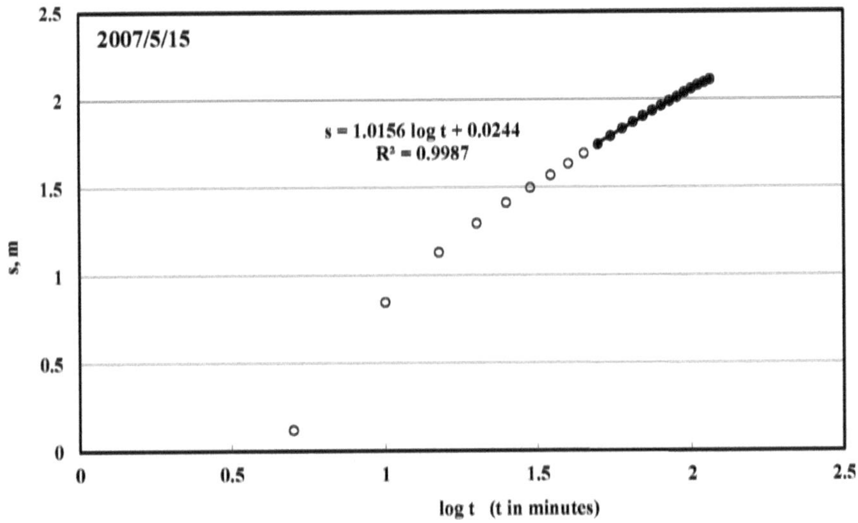

Fig. 5.4 The Jacob plot of drawdown data observed during the well test on May 15, 2007. From Kuo 2022 with permission of Groundwater, National Ground Water Association. Study of Water Relative Permeability in Fractures Using Well Tests and Radon: Gas Bubbles Effect. *Ground Water* 60(4):510–517

2008 M_W 5.4 Antung earthquake. As shown in Fig. 5.5b, the aquifer transmissivity decreased from the median (0.0349 m^2/min) to a minimum of 0.0214 m^2/min prior to the 2008 Antung earthquake. Likewise, as shown in Fig. 5.5a, the concentration of groundwater radon decreased from the median (705 pCi/L) to a minimum of 480 pCi/L prior to the 2008 Antung earthquake. The presence of gas bubbles induces both anomalies in groundwater-radon concentration and aquifer transmissivity precursory to the 2008 Antung earthquake. The agreement between both anomalies of groundwater radon and aquifer transmissivity provides strong field evidence regarding the development of gas bubbles in water-saturated cracked rock precursory to the 2008 Antung earthquake.

The box-and-whisker plot is used for statistical analysis of field data in groundwater radon and aquifer transmissivity. The box-and-whisker plot is shown on the right-hand side in Fig. 5.5a and b. The median (50th percentile, Q2) is shown as a center bar and the quartiles (25th and 75th percentiles, Q1 and Q3) as a box. The whiskers (upper and lower) cover all but the most extreme values in the data set. The lower whiskers are used to define the threshold values for identifying the anomalous declines in groundwater radon and aquifer transmissivity, which are also marked as the red bold lines in Fig. 5.5a and b. The anomalies in groundwater radon and aquifer transmissivity prior to the 2008 M_W 5.4 Antung earthquake can be clearly identified, respectively, using the threshold values (625 pCi/L and 0.0288 m^2/min).

Identifying precursors prior to earthquakes is a long-sought goal. Gas bubbles induce anomalous declines in both groundwater radon and aquifer transmissivity. Aquifer transmissivity can complement groundwater radon as an earthquake

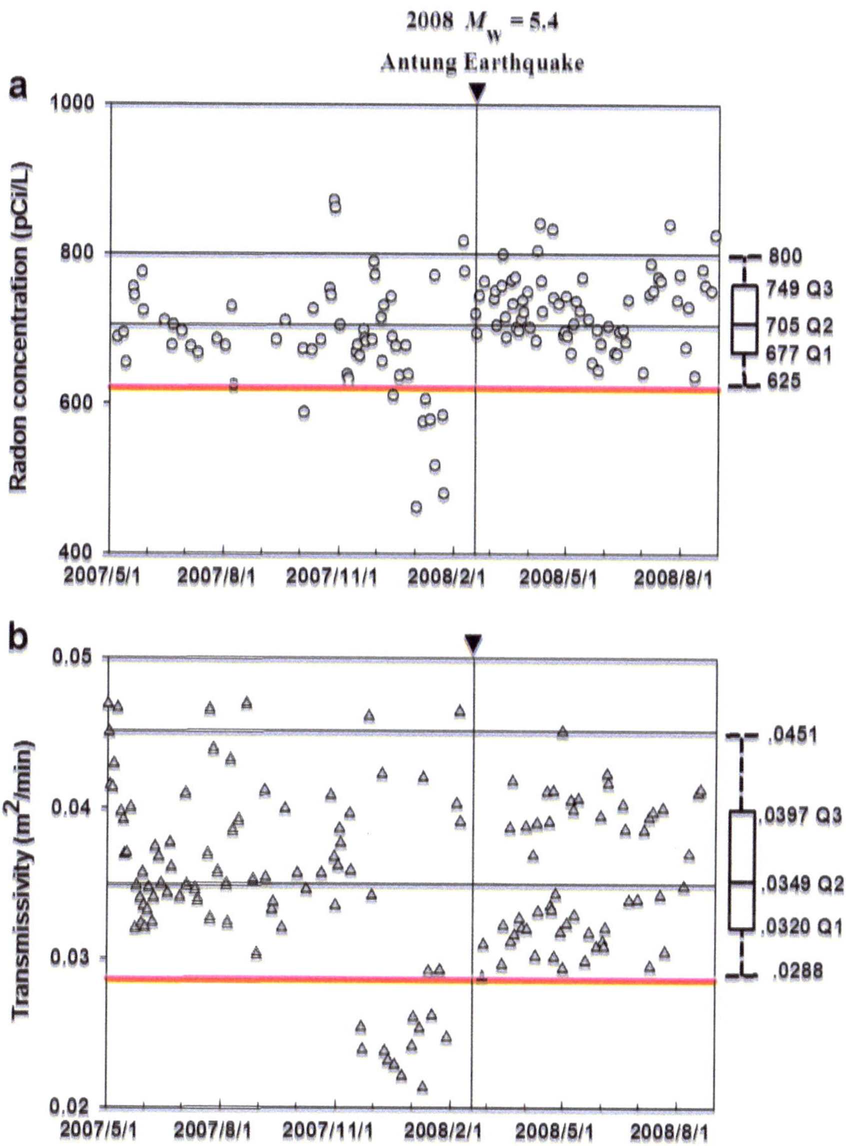

Fig. 5.5 Anomalous declines in (a) groundwater radon and (b) aquifer transmissivity precursory to the 2008 Antung earthquake. From Kuo 2022 with permission of *Groundwater*, National Ground Water Association

precursor. Here, gas bubbles induce two detectable earthquake precursors, groundwater radon and aquifer transmissivity. It is fortunate to discover that aquifer transmissivity can complement groundwater radon as an earthquake precursor (Kuo 2022). Simultaneous monitoring of aquifer transmissivity and groundwater radon is highly recommended for future studies in earthquake precursors.

5.4 Confirmation of In-Situ Volatilization Mechanism: Ruling Out Hypothesis of Groundwater Mixing*

* Kuo 2022 with permission of *Groundwater*, National Ground Water Association.

Figure 5.5 gives an idea for the development of gas bubbles prior to the 2008 M_w 5.4 Antung earthquake. Figure 5.5a reveals the presence of gas bubbles precursory to the 2008 Antung earthquake, which induces radon volatilization and causes anomalous declines in groundwater radon. Figure 5.5b also indicates the presence of gas bubbles precursory to the 2008 Antung earthquake, which reduces hydraulic conductivity and causes anomalous declines in aquifer transmissivity. Groundwater radon and aquifer transmissivity are two precursors having a common effect due to gas bubbles.

The radon anomalies observed at Antung well D1 can be explained by the in-situ volatilization mechanism. Nonetheless, groundwater mixing is another possible process to explain these precursory anomalies. A significant advance in this chapter is to decipher that the in-situ volatilization mechanism is responsible for the radon anomalies observed at Antung well D1. Groundwater mixing cannot explain the anomaly in aquifer transmissivity observed in Fig. 5.5b. Therefore, these anomalies are not caused by the groundwater mixing process. It is important to confirm that in-situ radon-volatilization is the mechanism responsible for the radon anomalies. With this confirmation, the radon anomalies observed at Antung well D1 can then be applied to estimate crustal strain quantitatively.

5.5 Favorable Geological Conditions for In-Situ Gas-Bubble Development

What are favorable geological conditions where gas bubbles can consistently develop precursory to local main earthquakes? The geology near Antung hot spring gives clues to the above question (Chen and Wang 1996). The Antung D1 well is situated in a small low-porosity fractured aquifer surrounded by a ductile mudstone, where groundwater recharge is limited. As regional stress increases, micro-cracks in brittle rock could form at a faster rate than the recharge rate of groundwater. Gas bubbles could develop precursory to a nearby earthquake in a small low-porosity fractured aquifer with undrained conditions. The gas bubbles induced anomalous decreases in both aquifer transmissivity and groundwater-radon concentration precursory to the 2008 M_w 5.4 Antung earthquake. Both the precursory data of aquifer transmissivity and groundwater-radon concentration are essential to gain an improved understanding of the precursory process of gas-bubble development in a small fractured aquifer with undrained conditions.

Recent global devastating events in the subduction zone, such as 2022 Taiwan Chihshang, 2022 Taiwan Kuanshan, 2018 Indonesia Sulawesi Island and 2011 Japan Tohoku earthquakes, occurred without warning. Recurrent radon anomalous declines consistently recorded precursory to the nearby main earthquakes are encouraging. Large main earthquakes to occur on the Longitudinal Valley Fault can be predicted a couple of months in advance. Likewise, prediction and warning of future destructive megathrust earthquakes in the subduction zone are possible months ahead of quake occurrence. The findings in this chapter can be globally applicable to subduction zones with similar tectonic settings and physical–chemical relationships.

The recurrent radon anomalies consistently observed at Antung well D1 suggest that the groundwater radon, when observed at suitable sites, can be a sensitive tracer for strain changes in crust associated with earthquake occurrences. A small low-porosity fractured aquifer with undrained conditions near an active fault can be an ideal natural strainmeter to site a radon-monitoring well for earthquake warnings. To corroborate the above hypothesis, the next chapter will present the results of monitoring groundwater radon at another site with similar geological conditions, Paihe Spring.

References

Brace WF, Paulding BW Jr, Scholz CH (1966) Dilatancy in the fracture of crystalline rocks. J Geophys Res 71(16):3939–3953

Chen CY, Horne RN, Fourar M (2004) Experimental study of liquid-gas flow structure effects on relative permeabilities in a fracture. Water Resour Res 40:W08301

Chen WS, Wang Y (1996) Geology of the coastal range, eastern Taiwan. Geology of Taiwan 7

Cooper HH, Jacob CE (1946) A generalized graphical method for evaluating formation constants and summarizing well field history. Am Geophys Union Trans 27:526–534

Erees FS et al (2007) Radon concentrations in thermal waters related to seismic events along faults in the Denizli Basin, Western Turkey. Radiat Meas 42:80–86

Hauksson E (1981) Radon content of groundwater as an earthquake precursor: evaluation of worldwide data and physical basis. J Geophys Res: Solid Earth 86(B10):9397–9410

Hauksson E, Goddard JG (1981) Radon earthquake precursor studies in Iceland. J Geophys Res: Solid Earth 86(B8):7037–7054

Hsu TL (1962) Recent faulting in the Longitudinal Valley of eastern Taiwan. Mem Geol Soc China 1:95–102

Kuo MCT (2022) Study of water relative permeability in fractures using well tests and radon: gas bubbles effect. Ground Water 60(4):510–517

Kuo MCT et al (2006) A mechanism for anomalous decline in radon precursory to an earthquake. Groundwater 44(5):642–647

Kuo T (2014) Correlating precursory declines in groundwater radon with earthquake magnitude. Ground Water 52(2):217–224

Kuo T, Chen W, Ho C (2018) Anomalous decrease in groundwater radon before 2016 Mw 6.4 Meinong earthquake and its application in Taiwan. Appl Radiat Isot 136:68–72

Kuo T et al (2011) Correlating recurrent radon precursors with local earthquake magnitude and crust strain near the Chihshang fault of eastern Taiwan. Nat Hazards 59:861–869

Kuo T et al (2013) Concurrent concentration declines in groundwater-dissolved radon, methane and ethane precursory to 2011 MW 5.0 Chimei earthquake. Radiat Meas 58:121–127

Kuo T et al (2017) A stress condition in aquifer rock for detecting anomalous radon decline precursory to an earthquake. Pure Appl Geophys 174:1291–1301

Morales-Simfors N, Wyss RA, Bundschuh J (2020) Recent progress in radon-based monitoring as seismic and volcanic precursor: a critical review. Crit Rev Environ Sci Technol 50(10):979–1012

Noguchi M, Wakita H (1977) A method for continuous measurement of radon in groundwater for earthquake prediction. J Geophys Res 82(8):1353–1357

Nur A (1972) Dilatancy, pore fluids, and premonitory variations of ts/tp traval times. Bull Seismol Soc Am 62(5):1217–1222

Persoff P, Pruess K (1995) Two-phase flow visualization and relative permeability measurement in natural rough-walled rock fractures. Water Resour Res 31:1175–1186

Roeloffs E (1999) Radon and rock deformation. Nature 399:104–105

Scholz CH, Sykes LR, Aggarwal YP (1973) Earthquake prediction: A physical basis. Science 181(4102):803–810

Shapiro MH, Melvin JD, Tombrello TA (1980) Automated radon monitoring at a hard-rock site in the southern California transverse ranges. J Geophys Res 85:3058–3064

Tarakçı M et al (2014) Investigation of the relationships between seismic activities and radon level in western Turkey. Appl Radiat Isot 83:12–17

Theis CV (1935) The relation between the lowering of piezometric surface and the rate of the duration of discharge of well using groundwater storage. Trans Am Geophys Union 16:519–524

Wakita H et al (1980) Radon anomaly: a possible precursor of the 1978 Izu-Oshima-kinkai earthquake. Science 207(4433):882–883

Wakita H (1996) Geochemical challenge to earthquake prediction. Proc Natl Acad Sci USA 93:3781–3786

Weigel F (1978) Radon. Chemiker-Zeitung 102:287–299

Wu YM et al (2006) Coseismic versus interseismic ground deformations, fault rupture inversion and segmentation revealed by 2003 Mw 6.8 Chengkung earthquake in eastern Taiwan. Geophys Res Lett 33:L02312.

Zoback MD, Byerlee JD (1975) The effect of cyclic differential stress on dilatancy in westerly granite under uniaxial and triaxial conditions. J Geophys Res 80(11):1526–1530

Chapter 6
Anomalous Radon Declines in Small Unconfined Aquifers: Corroboration of Favorable Geological Conditions

Abstract In Chaps. 4 and 5, favorable geological conditions, a small low-porosity brittle aquifer surrounded by a ductile formation, were attributed to catch the recurrent radon anomalous declines at Antung hot spring precursory to the local main earthquakes. In this chapter, we will test the above hypothesis by monitoring groundwater radon at the Paihe Spring with similar geological conditions at Antung. The other hard-rock site, Byakko Spring in Japan, will also be discussed. Both springs, Paihe and Byakko, are small unconfined aquifers. The groundwater-radon concentration in an unconfined aquifer is markedly affected by precipitation. Unlike a confined aquifer, an unconfined aquifer can be used for earthquake warnings only in the dry season.

Keywords Unconfined aquifers · Groundwater Radon · Dissolved gases · Precursors

6.1 Introduction

The local geological conditions were attributed for catching the recurrent radon anomalous declines precursory to the nearby large earthquakes near well D1 in the Antung hot spring (Kuo et al. 2006). The Antung hot spring is a small low-porosity brittle aquifer surrounded by a ductile formation in undrained conditions. The main objective of this chapter is to test the above hypothesis by monitoring groundwater radon at other sites with similar geological conditions. The Paihe Spring is also a small brittle limestone aquifer surrounded by ductile formations. The Paihe Spring is an unconfined aquifer. Aquifer recharge is limited in the dry season and can be considered in undrained conditions.

The other hard-rock site discussed in this chapter is Byakko spring in Japan. Precursory to the 1984 M 6.8 Western Nagano earthquake, anomalous declines in groundwater-dissolved gas were observed at the Byakko spring. Anomalous declines in groundwater-dissolved gas recorded by Sugisaki and Sugiura (1986) precursory to

M. T. Kuo, *Groundwater Radon in the Taiwan Subduction Zone*, Advances in Geological Science,
https://doi.org/10.1007/978-981-99-5350-9_6

the 1984 Western Nagano earthquake corroborated the mechanism of in-situ radon-volatilization proposed in Chap. 4. The 1984 anomaly recorded at the Byakko spring also confirmed the favorable geological conditions outlined in Chap. 5 for in-situ gas-bubble development.

6.2 Geological Settings of Paihe Spring Versus Antung Hot Spring*

*Kuo et al. (2018) with permission of Applied Radiation and Isotopes.

Based on the local geological conditions near the Antung hot spring in eastern Taiwan, the Paihe Spring in southwestern Taiwan with similar geological conditions was selected to corroborate the favorable geological conditions for consistently catching precursory declines in groundwater radon. The Paihe Spring is an unconfined small limestone aquifer.

Observation of groundwater radon was initiated at the Paihe Spring in November 2009. Radon anomalous declines were recorded at the Paihe Spring P1 prior to the 2010 Jiasian and 2016 Meinong earthquakes. Here, the geological conditions favorable for in-situ gas-bubble development at a confined aquifer are confirmed also valid at an unconfined aquifer. Figure 6.1 shows the location of the Paihe Spring P1 and the epicenters of the 2010 M_W 6.3 Jiasian, and 2016 M_W 6.4 Meinong earthquakes near the area southeast of Chiayi. The 1964 M_L 6.3 Paihe earthquake was the strongest in the Chiayi-Tainan area between 1964 and 2009.

Figure 6.2 shows that the Paihe Spring is situated in a brittle limestone aquifer surrounded by the ductile Yunshuichi Formation consisting of shale and sandy shale. The limestone is essentially massive reef corals. The limestone forms an isolated, flat-topped peak called Chentou Mountain, which is about 646 m above sea level. The limestone lentils strike between 30° and 50° to the northeast and dip between 50° and 60° to the northwest. The limestone extends about 900 m with an average thickness of 40 m (Yen and Chang 1949). The Paihe Spring is a small brittle aquifer isolated and surrounded by a ductile mudstone. Precipitation infiltrates into the limestone lentils. The aquifer recharge only occurs during the rainy season. During the dry season (from the end of November to the end of March next year), the unconfined limestone aquifer at Chentou Mountain can be regarded in undrained conditions. Environmental factors particularly rainfall affect the radon concentration of groundwater samples taken from unconfined aquifers. For an earthquake precursor study, it is important to ensure that the radon concentration of groundwater samples taken from an unconfined aquifer is not affected by rainfall.

Figure 6.3 shows that the Antung hot spring is situated in an exotic brittle tuffaceous sandstone block surrounded by a ductile mudstone of the Paliwan Formation and the Lichi mélange. The exotic block extends about 1 or 2 km (Chen and Wang 1996). The Antung hot spring is a small brittle aquifer isolated and surrounded by a ductile mudstone. Well D1 is not artesian and aquifer recharge is weak. The aquifer

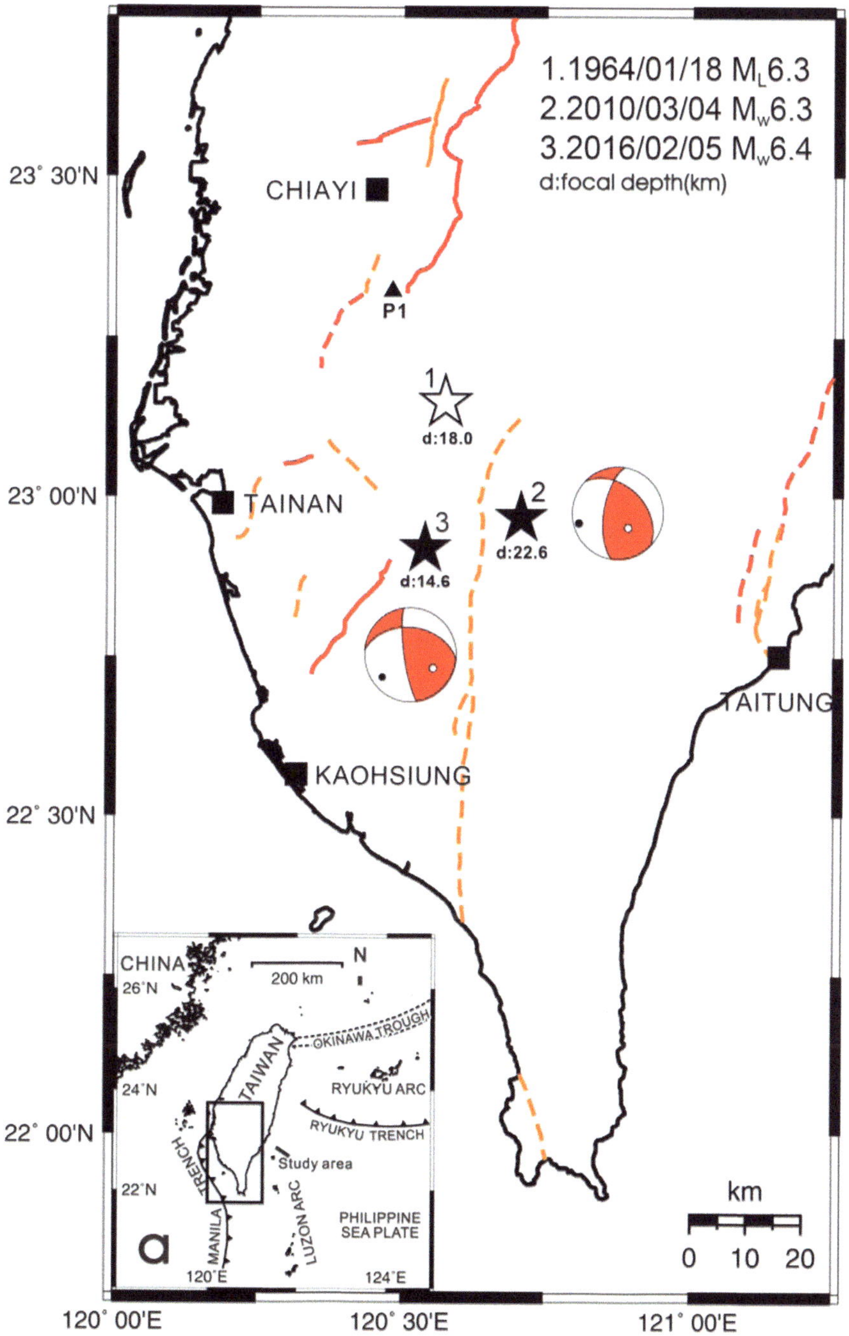

Fig. 6.1 Map of the epicenters of the earthquakes that occurred on January 8, 1964, March 4, 2010, and February 5, 2016, in southwestern Taiwan (Kuo et al. 2018 with permission of Applied Radiation and Isotopes) **a** Map of Taiwan. **b** The study area (open star: 1964 mainshock, filled stars: 2010 Jiasian and 2016 Meinong mainshocks, filled triangle: Paihe Spring P1)

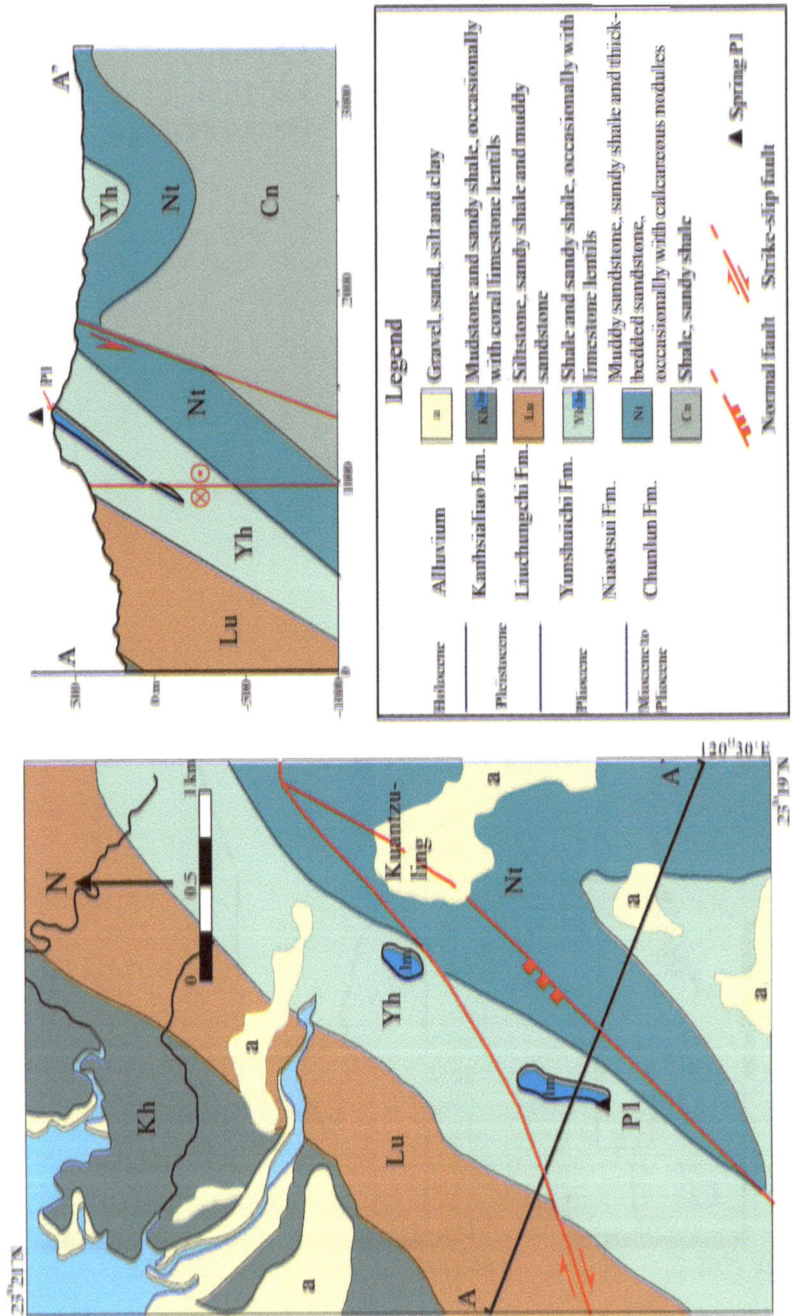

Fig. 6.2 Geological map and cross section near the Paihe Spring (P1) (Kuo et al. 2018 with permission of Applied Radiation and Isotopes)

near well D1 can be regarded in undrained conditions. The hot spring is formed nearby an eastward-dipping, Yongfeng Fault. Unlike the Paihe Spring, the Antung hot spring is a confined aquifer. Rainfall does not affect the radon concentration of groundwater samples taken from well D1.

In a small brittle aquifer under undrained conditions (Antung well D1 or Paihe Spring P1), the development of new cracks in the aquifer rock could occur at a rate faster than the recharge of pore water prior to an earthquake. In a small brittle aquifer with undrained conditions, gas saturation and two phases (vapor and liquid) could develop in the rock cracks prior to an earthquake (Nur 1972; Scholz et al. 1973; Kuo 2022). Meanwhile, the radon in groundwater volatilizes and partitions into the gas phase and the concentration of radon in groundwater decreases. The above mechanism is also referred to as "in-situ radon volatilization" (Kuo et al. 2006). The geological conditions favorable for the development of gas-bubble in a confined aquifer (Antung well D1) are also applicable to an unconfined aquifer (Paihe Spring P1). Table 6.1 summarizes the geological conditions at Antung well D1 and Paihe Spring P1.

6.3 Anomalous Radon Declines Before the 2010 Mw 6.3 Jiasian and 2016 Mw 6.4 Meinong Earthquakes: Paihe Spring, Taiwan*

*Kuo et al. (2018) with permission of Applied Radiation and Isotopes.

The monitoring of groundwater radon at the Paihe Spring P1 began in November 2009. Radon anomalous declines were recorded at the Paihe Spring P1 prior to the 2010 M_w 6.3 Jiasian and 2016 M_w 6.4 Meinong earthquakes. The epicenters of the 2010 Jiasian and 2016 Meinong earthquakes are located 46 km and 45 km from the Paihe Spring P1, respectively. Figures 6.4 and 6.5 show the two radon anomalies observed at the Paihe Spring P1. Specifically, the concentration of groundwater radon decreased from background levels of 144 ± 7 and 137 ± 8 pCi/L to minima of 104 ± 8 and 97 ± 9 pCi/L prior to the 2010 Jiasian and 2016 Meinong earthquakes, respectively. Radon-concentration errors are ± 1 standard deviation after simple averaging of triplicates. The Paihe Spring is an unconfined limestone aquifer. Rainfall is the main environmental factor that significantly affects the radon concentration of groundwater samples. Fortunately, the radon anomalies shown in Figs. 6.4 and 6.5 were recorded in the dry season, when there was no rainfall affecting the radon concentration of groundwater samples. For an earthquake precursory study in an unconfined aquifer such as the Paihe Spring P1, it is important to make sure that the radon concentration of groundwater samples are taken during the dry season.

Figures 6.4 and 6.5 show the anomalies observed at the Paihe Spring P1 follow the three-stage pattern like the anomalies observed at Antung well D1. The sequence of events for radon anomalies prior to the 2010 M_w 6.3 Jiasian and 2016 M_w 6.4 Meinong earthquakes can be characterized by three stages. Stage 1 is buildup of

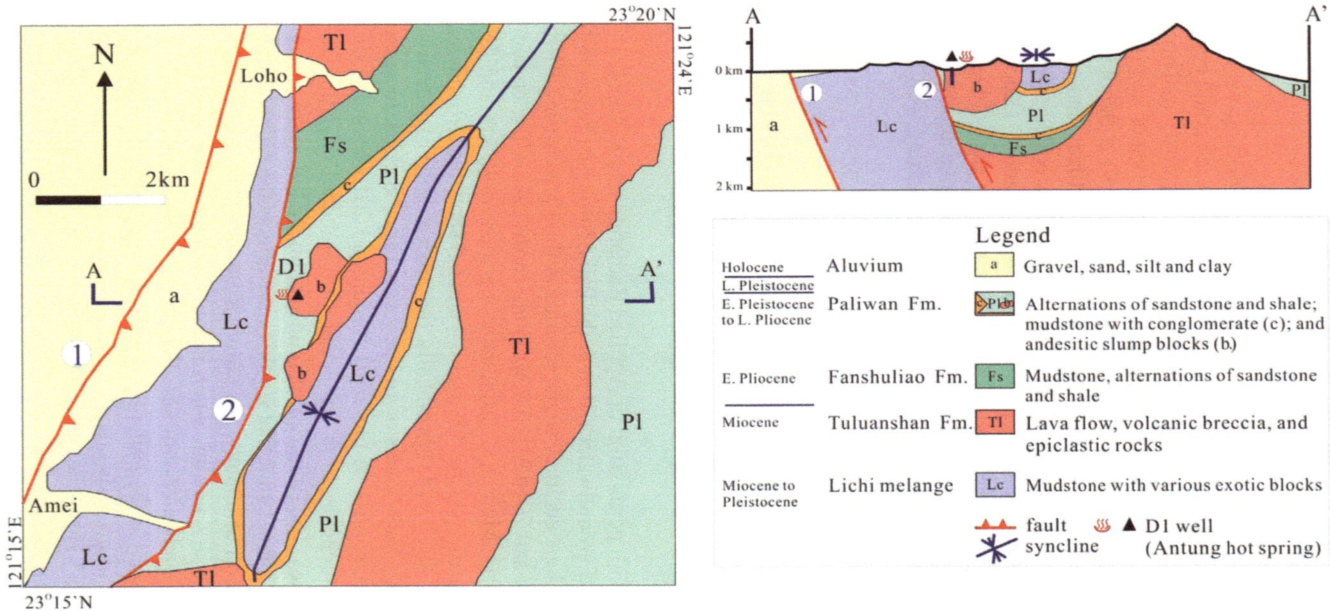

Fig. 6.3 Geological map and cross section near the radon-monitoring well D1 in the area of Antung hot spring. (Kuo et al. 2018 with permission of Applied Radiation and Isotopes) (①: Chihshang, or, Longitudinal Valley Fault, ②: Yongfeng Fault)

Table 6.1 Summary of geological conditions at Antung well D1and Paihe Spring P1

Site	Aquifer rock	Aquifer type	Aquifer size (length, width, thickness)	Aquifer recharge strength	Surroundings rock
Antung well D1	Brittle Andesite	Confined	Small (1500 m, 500 m, 15 m)	Weak	Ductile Mudstone
Paihe Spring P1	Brittle Limestone	Unconfined	Small (900 m, 300 m, 40 m)	Weak in Dry Season	Ductile Shale

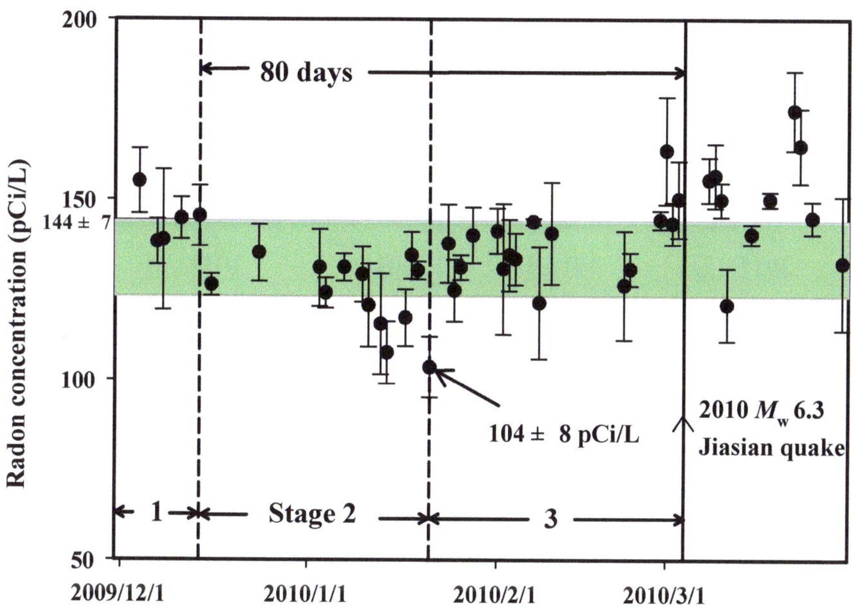

Fig. 6.4 Observed radon anomalies at the Paihe Spring (P1) prior to 2010 Jiasian earthquake. (Kuo et al. 2018 with permission of Applied Radiation and Isotopes) Green rectangles show radon concentration between the mean radon concentration and three standard deviations below the mean. Stage 1 is buildup of elastic strain. Stage 2 is the development of cracks. Stage 3 is the influx of groundwater

elastic strain. During Stage 1, the radon concentration in groundwater is fairly stable. Stage 2 is the development of cracks and gas bubbles in the brittle aquifer. During Stage 2, the radon in groundwater volatilizes into the gas bubbles. The radon concentration of groundwater starts to decrease and reaches a minimum value. Stage 3 is influx of groundwater. Stage 3 starts at the point of minimum radon concentration. During Stage 3, the radon concentration in groundwater increases and recovers to the previous background level before the earthquake. The rule to define stages 1, 2, and 3 in Figs. 6.4 and 6.5 is based on the trend regarding the temporal behavior of radon concentration. During Stage 1, radon concentration is fairly stable. During Stage 2,

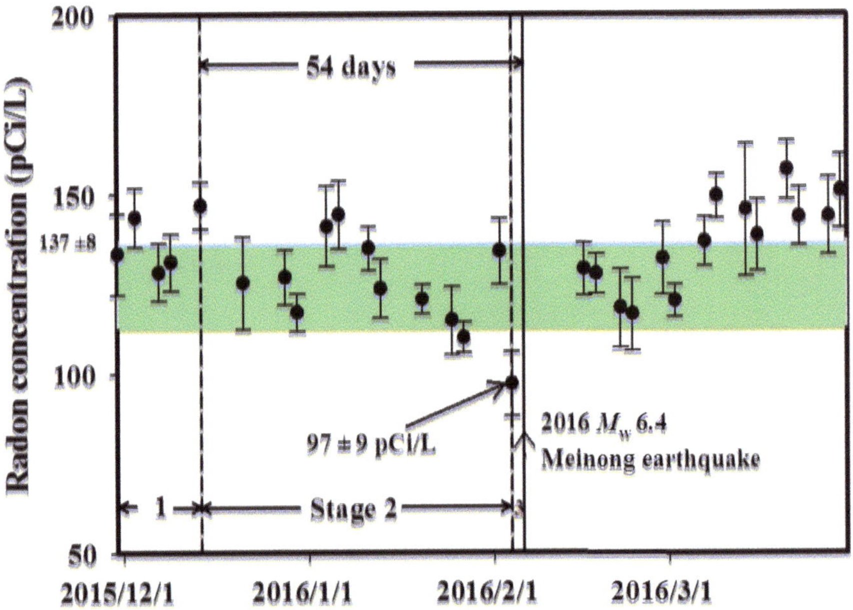

Fig. 6.5 Observed radon anomalies at the Paihe Spring (P1) prior to 2016 Meinong earthquake. (Kuo et al. 2018 with permission of Applied Radiation and Isotopes) Green rectangles show radon concentration between the mean radon concentration and three standard deviations below the mean. Stage 1 is buildup of elastic strain. Stage 2 is the development of cracks. Stage 3 is the influx of groundwater

radon concentration decreases. During Stage 3, radon concentration increases. The three-stage pattern of groundwater-radon anomalies shown in Figs. 6.4 and 6.5 can be explained by the in-situ radon-volatilization mechanism (Kuo et al. 2006). The three-stage pattern recognized in the radon anomalies before the 2010 Jiasian and 2016 Meinong earthquakes is valuable for early warning of local large earthquakes.

An anomaly at the Paihe Spring P1 is defined as a significant deviation from the mean value, or, three standard deviations below the mean value (Zmazek et al. 2006). As shown in Figs. 6.4 and 6.5, the mean radon concentration and associated standard deviation of the background level are calculated using radon data from Stage 1 (144 ± 7 and 137 ± 8 pCi/L for the 2010 Jiasian and 2016 Meinong earthquakes, respectively). The green rectangles in Figs. 6.4 and 6.5 are used to identify radon anomalies at the Paihe Spring P1, which show radon concentration between the mean radon concentration and three standard deviations below the mean.

The precursor time for radon anomalous decline is defined as the time interval between the start of Stage 2 when the radon concentration starts to decline and the time of occurrence of the earthquake. As shown in Figs. 6.4 and 6.5, the precursor times for radon anomalies were 80 and 54 days prior to the 2010 M_w 6.3 Jiasian and 2016 M_w 6.4 Meinong earthquakes, respectively. Based on focal mechanisms in Fig. 6.1, the 2010 M_w 6.3 Jiasian, and 2016 M_w 6.4 Meinong earthquakes occurred on different faults. Therefore, Figs. 6.4 and 6.5 also imply that the radon-decline behavior and precursory time are the characteristics of the causative fault.

Figure 6.1 shows that the Paihe Spring P1 is located 46 km and 45 km from the epicenters of the 2010 M_w 6.3 Jiasian and 2016 M_w 6.4 Meinong earthquakes, respectively. Figures 6.4 and 6.5 show radon anomalies recurrently observed at the Paihe Spring P1 prior to the 2010 Jiasian and 2016 Meinong earthquakes, respectively. Specifically, the concentration of groundwater radon decreased from background levels of 144 ± 7 and 137 ± 8 pCi/L to minima of 104 ± 8 and 97 ± 9 pCi/L prior to the 2010 Jiasian and 2016 Meinong earthquakes, respectively. The above radon anomalies observed at the Paihe limestone spring corroborated that the geological conditions favorable for in-situ gas-bubble development at a confined aquifer are also valid at an unconfined aquifer. The unconfined limestone aquifer at Paihe is in undrained conditions during the dry season and can be used for earthquake warnings.

Figure 6.6 shows the groundwater-radon concentration versus precipitation in an unconfined limestone aquifer in Kenting, Taiwan. Unlike in a confined aquifer, the groundwater-radon concentration in an unconfined aquifer is markedly affected by precipitation. For an earthquake-precursor study, it is essential that the radon concentration of groundwater samples is not affected by rainfall. While a confined aquifer can be used all year round, an unconfined aquifer is used for earthquake warning only during the dry season.

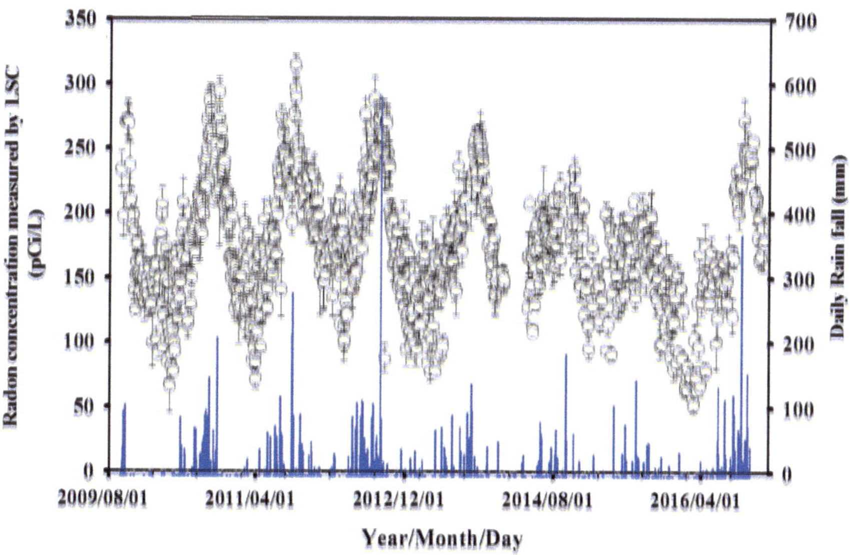

Fig. 6.6 Groundwater-radon concentration versus precipitation in an unconfined limestone aquifer at Kenting, Taiwan

6.4 Anomalous Gas Declines Before the 1984 M 6.8 Western Nagano Earthquake: Byakko Spring, Japan

Sugisaki and Sugiura (1986) observed simultaneous concentration declines in groundwater-dissolved gases at Byakko Spring, precursory to the 1984 M 6.8 Western Nagano earthquake. Figure 6.7 shows the location of Byakko Spring and epicenter of the 1984 M 6.8 Western Nagano earthquake. The M 6.8 inland earthquake occurred on September 14, 1984, with a focal depth of 3.8 km in central Japan. The 1984 Western Nagano earthquake caused a large landslide and resulted in a toll of 29 lives, in spite of the fact that it occurred in a sparsely populated mountainous area.

The Byakko Spring is located in an area of Cretaceous granite and 50 km from the epicenter of the 1984 Western Nagano earthquake. The composition of major dissolved gases in groundwater at the Byakko Spring consists of 84.2% nitrogen, 13.9% methane, 0.658% argon, and 0.0645% helium (Sugisaki and Sugiura 1986). Gas bubbles at the Byakko Spring are collected continuously with a funnel installed upside down in the spring water. The sample gases are directly introduced into the gas chromatograph on site and analyzed every 6 h.

Figure 6.8 shows the temporal variations in volume ratio of He/Ar, N_2/Ar, and CH_4/Ar before and after the 1984 M 6.8 Western Nagano earthquake (Sugisaki and Sugiura 1986). The three ratios show synchronous variations, increasing with relatively small fluctuations until April 1984. Then, the three ratios all dropped abruptly and reached a minimum at the end of July 1984. About 40 days after the

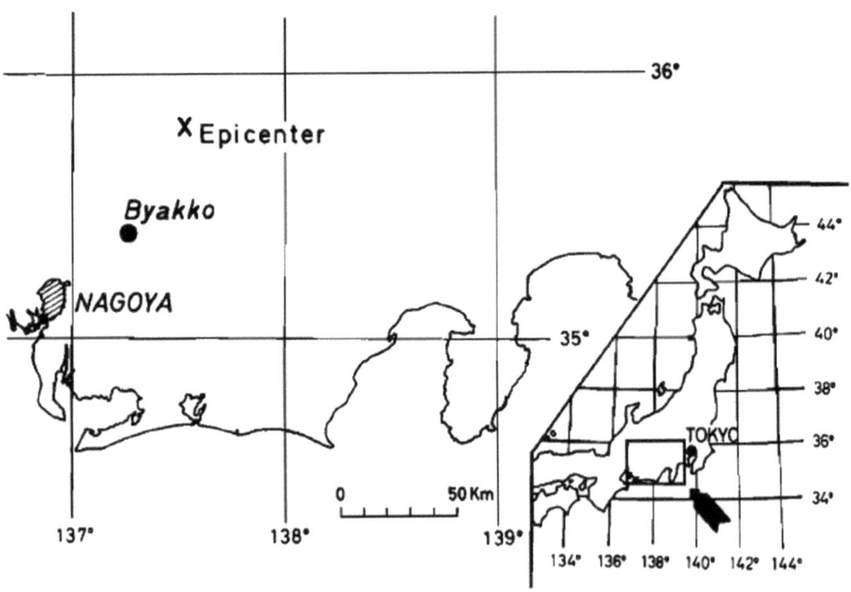

Fig. 6.7 Location of Byakko Spring and epicenter of the 1984 M 6.8 Western Nagano earthquake (From Sugisaki and Sugiura 1986)

minimum, the 1984 M 6.8 Western Nagano earthquake occurred on September 14, 1984.

As shown in Fig. 6.8, the anomalous decreases in the three ratios observed at the Byakko Spring follow the same v-shaped trend as those observed at Paihe and Antung. The precursory time estimated from Fig. 6.8 is about 150 days. The gas anomaly recorded at Byakko precursory to the 1984 Western Nagano earthquake is striking. The anomalous decreases in the three ratios corroborated the in-situ volatilization mechanism. The Byakko Spring is a hard-rock granite site. The 1984 anomaly recorded at the Byakko Spring also supported that a low-porosity fractured aquifer can be a favorable geological condition for in-situ gas-bubble development.

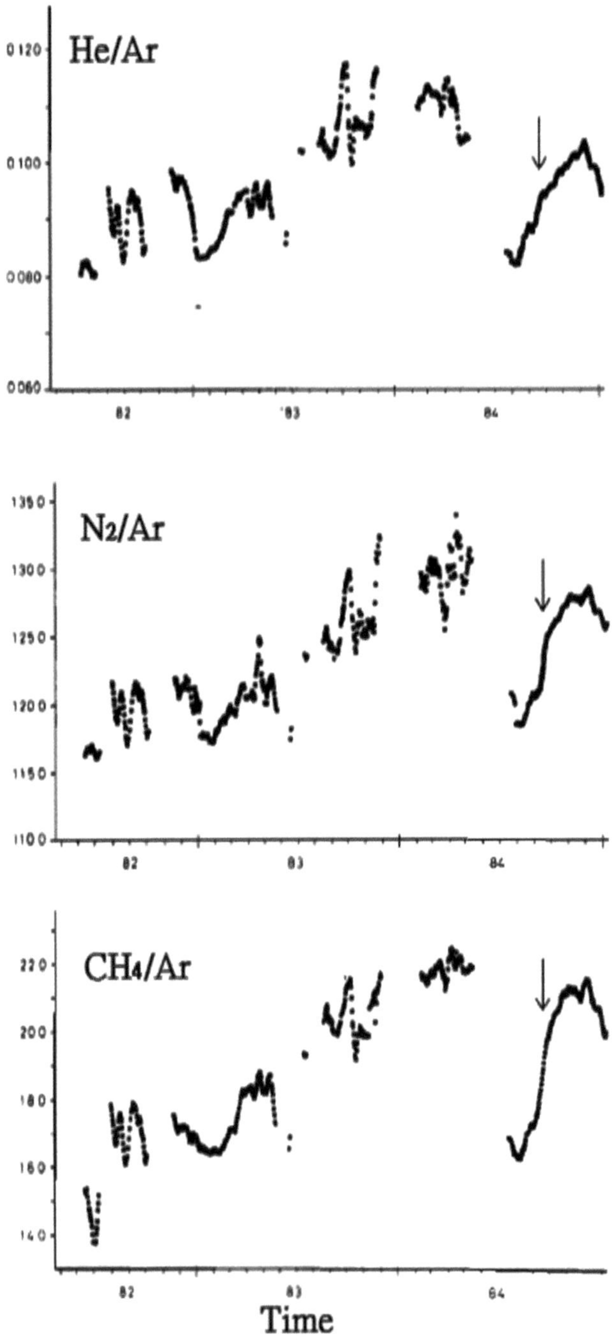

Fig. 6.8 Temporal variations in volume ratio of He/Ar, N_2/Ar, and CH_4/Ar before and after the 1984 M 6.8 Western Nagano earthquake. Arrows indicate the occurrence of the earthquake (From Sugisaki and Sugiura 1986)

References

Chen WS, Wang Y (1996) Geology of the coastal range, eastern Taiwan. In: Geology of Taiwan 7

Kuo MCT (2022) Study of water relative permeability in fractures using well tests and radon: gas bubbles effect. Ground Water 60(4):510–517

Kuo T, Chen W, Ho C (2018) Anomalous decrease in groundwater radon before 2016 Mw 6.4 Meinong earthquake and its application in Taiwan. Appl Radiat Isot 136:68–72

Kuo MCT et al (2006) A mechanism for anomalous decline in radon precursory to an earthquake. Ground Water 44(5):642–647

Nur A (1972) Dilatancy, pore fluids, and premonitory variations of ts/tp traval times. Bull Seismol Soc Am 62(5):1217–1222

Scholz CH, Sykes LR, Aggarwal YP (1973) Earthquake prediction: a physical basis. Science 181(4102):803–810

Sugisaki R, Sugiura T (1986) Gas anomalies at three mineral springs and a fumarole before an inland earthquake, central Japan. J Geophys Res Solid Earth 91(B12):12296–12304

Yen TP, Chang LS (1949) The Kwantzuling limestone. Bull Geol Surv Taiwan 2:51–52

Zmazek B et al (2006) Radon in a thermal spring: Identification of anomalies related to seismic activity. Appl Radiat Isot 64(6):725–734

Chapter 7
Anomalous Radon Declines in Small Confined Aquifers: Evaluation of Global Data

Abstract In Chap. 6, we reviewed anomalous radon declines in small unconfined aquifers. In this chapter, we will review anomalous radon declines in small confined aquifers. Global data from Japan, Iceland, and Taiwan will be compared to each other. Radon anomalies will be evaluated in terms of geological conditions, monitoring methods, epicenter distance, radon-decline percent, precursory time, and earthquake magnitude. With the help of a case study at the Antung hot spring, this chapter will also illustrate the application of recurrent radon anomalies for forecasting local main earthquakes occurring on the Longitudinal Valley Fault in the Taiwan subduction zone.

Keywords Confined aquifers · Groundwater · Radon anomaly · Earthquake forecast · Taiwan subduction zone

7.1 Introduction

Worldwide data manifesting anomalous radon declines in confined aquifers are reviewed in this chapter. The data of anomalous declines in groundwater-radon concentration are valuable for the early warning of local disastrous earthquakes. The observed radon declines in confined aquifers can be applied to quantify the gas saturation and volumetric strain in the confined fractured aquifer through simple analytical models.

In this chapter, case studies from Japan, Iceland, and Taiwan are reviewed and compared with each other. The data from Taiwan, Japan, and Iceland are evaluated in terms of geological conditions, monitoring methods, epicenter distance, radon-decline percent, precursory time, and earthquake magnitude. With the help of a case study at the Antung hot spring in Taiwan, it is valuable to correlate radon decline and precursory time with earthquake magnitude for a given radon-monitoring site (well D1) and for a given active fault (Longitudinal Valley Fault). These correlations are useful for early warning of local large main earthquakes occurring on the Longitudinal Valley Fault.

M. T. Kuo, *Groundwater Radon in the Taiwan Subduction Zone*, Advances in Geological Science,
https://doi.org/10.1007/978-981-99-5350-9_7

7.2 Japan: Radon Anomaly Before the 1978 Izu-Oshima-Kinkai Earthquake (M 7.0)

Anomalous decrease in groundwater-radon concentration prior to the 1978 M 7.0 Izu-Oshima-Kinkai earthquake was recorded at the radon-monitoring station (SKE-1) near Nakaizu on the Izu Peninsula (Wakita et al. 1980). Figure 7.1 shows the epicenter of the main shock located at 34.8°N and 139.3°E near east of the Izu Peninsula, Japan. The distance between the epicenter and the radon-monitoring station (SKE-1) was about 25 km for the 1978 Izu-Oshima-Kinkai earthquake. Based on a seismological study (Shimazaki and Somerville 1978), the hypocenter was on an east-trending right-lateral strike-slip fault 17 km long, which was situated beneath the sea between the Izu Peninsula and Izu-Oshima Island. Ground displacement occurred along the western edge of the fault with a length of 3 km trending northwest–southeast on Izu Peninsula's east coast (Tsuneishi 1978). The maximum displacement of the fault was 1.3 m.

The radon anomaly precursory to the 1978 M 7.0 Izu-Oshima-Kinkai earthquake is well documented (Wakita et al. 1980; Wakita 1996). Figure 7.2a shows the radon anomaly observed at the SKE-1 station. The radon concentration at the SKE-1 monitoring well began to decrease from a background level of 483 ± 3 cpm in the middle of October 1977. The precursory time of the 1978 radon anomaly was about eighty days before the 1978 Izu-Oshima-Kinkai earthquake, which occurred on January 14, 1978. The radon concentration declined to a minimum level of 439 ± 7 cpm in

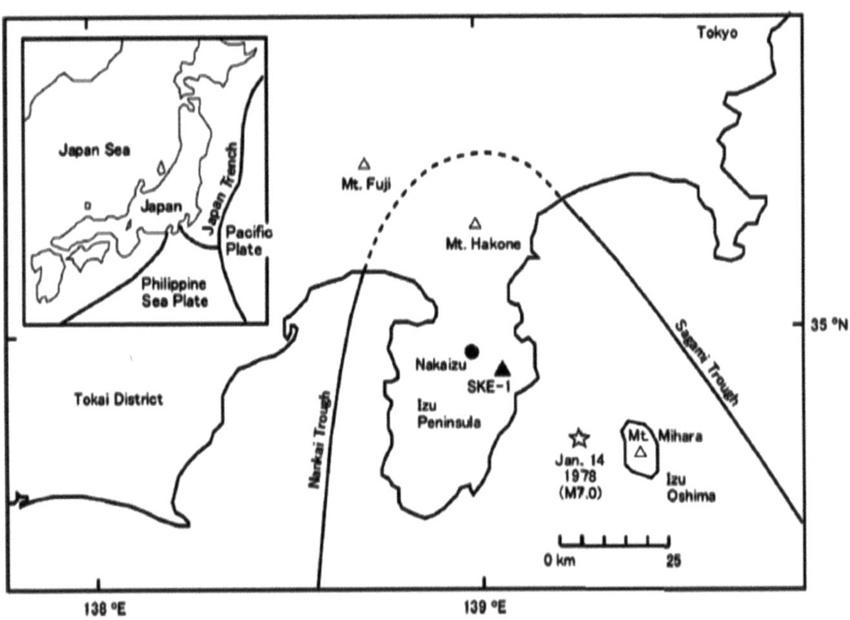

Fig. 7.1 Map of the Izu Peninsula and the surrounding area (From Tsunomori and Kuo 2010)

December 1977. Then, the concentration increased rapidly, reaching a higher level than the previous background level before the main shock.

The radon anomaly observed at the SKE-1 well precursory to the 1978 M 7.0 Izu-Oshima-Kinkai earthquake behaved similarly to that observed at the Antung D1 well precursory to the 2003 M_w 6.8 Chengkung earthquake, Taiwan (Kuo et al. 2006). The 20 km distance from the epicenter to the Antung D1 well for the 2003 M_w 6.8 Chengkung earthquake is close to the 25 km distance between the epicenter and the SKE-1 well for the 1978 Izu-Oshima-Kinkai earthquake. The radon concentration of groundwater started to decrease sixty-five days before the 2003 Chengkung earthquake. Similarly, about eighty days before the 1978 Izu-Oshima-Kinkai earthquake, the radon concentration started to decrease. Both the SKE-1 and Antung D1 wells are completed in fractured aquifers. Both studies (Wakita et al. 1980; Kuo et al. 2006) corroborate that under suitable geological conditions, radon declines in groundwater can be a sensitive tracer of strain changes in the crust preceding an earthquake.

The SKE-1 well is artesian with a total depth of 350 m. A continuous monitoring system was used to count radon concentration in groundwater at the SKE-1 well (Noguchi and Wakita 1977). A separation chamber in which radon is emanated from groundwater to the gaseous phase. The radon concentration in the gaseous phase equilibrates with that in the water phase. The measuring system consists of a ZnS scintillation chamber for alpha counting. Alpha activities of radon and its daughters (218Po and 214Po) in the scintillation chamber are counted and recorded continuously.

The SKE-1 well is situated in a fractured confined aquifer and located in the valley of an isolated mountainous area near Nakaizu in the Izu Peninsula. When aquifer recharge is weak, dilation of the rock mass may occur at a rate faster than the rate at which pore water can flow into the newly created pore volume when regional stress increases (Nur 1972; Scholz et al. 1973). Gas saturation and two phases (vapor and liquid) could develop in the rock cracks. Meanwhile, the dissolved radon escapes from groundwater into the gas phase and the radon concentration in groundwater decreases (Kuo et al. 2006).

The radon anomaly observed at the SKE-1 well prior to the 1978 Izu-Oshima-Kinkai earthquake follows a three-stage pattern like those observed at the Antung well D1. As shown in Fig. 7.2a, Stage 1 was from September 1, 1977, to October 24, 1977. During Stage 1, radon was fairly stable (around 483 $\pm$ 3 count/min) and there was an accumulation of tectonic strain, which produced a slow, steady increase of effective stress. About eighty days before the 1978 Izu-Oshima-Kinkai earthquake, the concentration of radon started to decrease and reached a minimum level of 439 $\pm$ 7 count/min during Stage 2 for a 60-day period. The aquifer was under undrained condition during Stage 2. Dilation of the rock mass occurred and gas saturation developed in micro-cracks. Groundwater radon volatilized into the gas phase during Stage 2. Stage 3 started at the point of minimum radon concentration. During Stage 3, water saturation in cracks began to increase with water influx. The radon concentration in groundwater recovered to the previous background level before the 1978 Izu-Oshima-Kinkai earthquake.

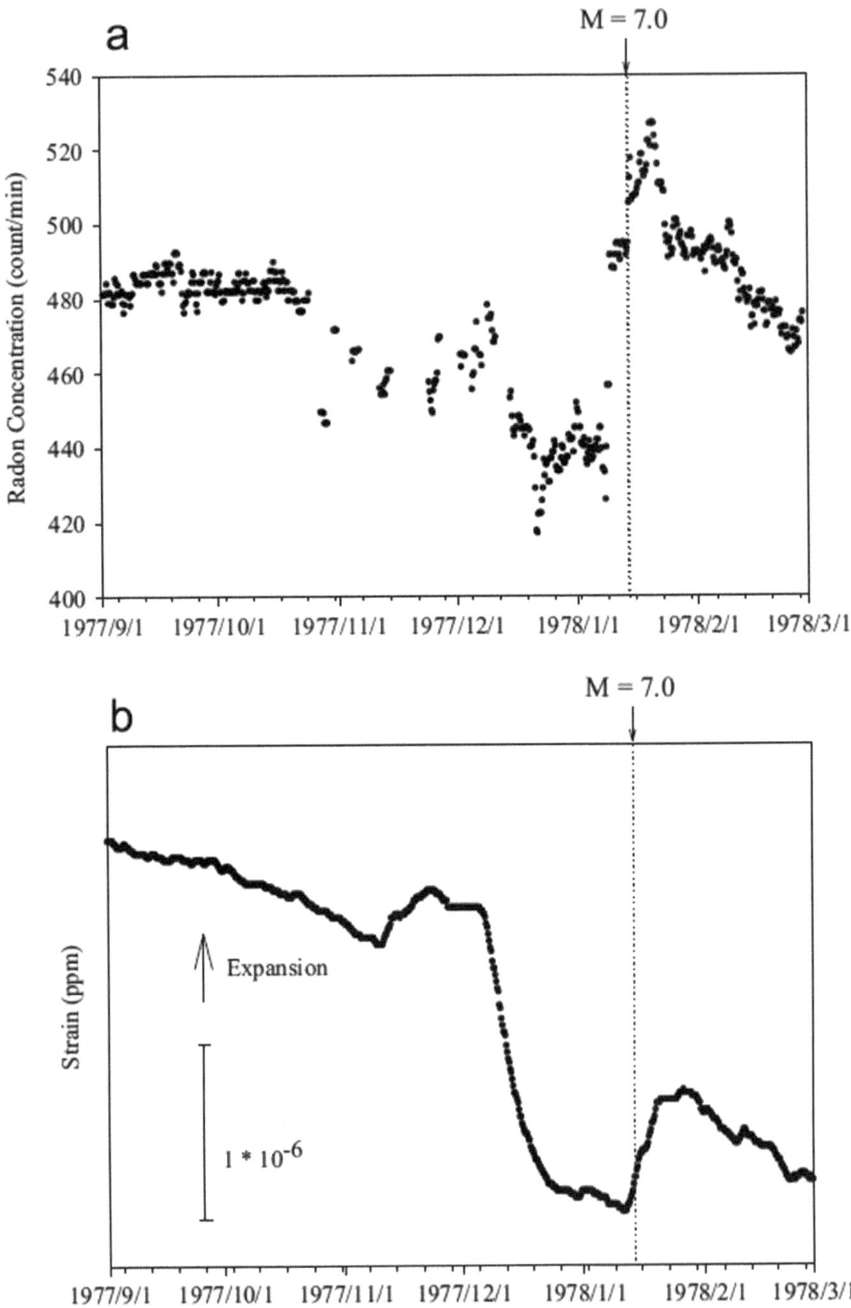

Fig. 7.2 Precursory changes of the 1978 Izu-Oshima-Kinkai earthquake (From Tsunomori and Kuo 2010). **a** Radon concentration changes observed at the SKE-1 well (350 m deep) with a distance from the epicenter (D) = 25 km. **b** Record of the volumetric strainmeter at Irozaki with D = 50 km

The physical mechanism of groundwater-radon volatilization can be applied to explain the anomalous declines in groundwater radon observed prior to the 1978 Izu-Oshima-Kinkai earthquake (Tsunomori and Kuo 2010; Wakita et al. 1980). For a confined aquifer with undrained conditions, in-situ radon-volatilization model, Eq. (7.1), correlates the observed decline in groundwater radon with the gas saturation developed in the fractured aquifer (Kuo et al. 2006).

$$C_0 = C_w(H \times S_g + 1) \tag{7.1}$$

where C_0 is initial radon concentration in groundwater precursory to radon decrease, pCi/L; C_w is the minimum radon concentration in groundwater observed during radon anomaly, pCi/L; S_g is gas saturation, fraction; H is Henry's coefficient for radon at aquifer temperature, dimensionless.

Equation (7.1) can be applied to quantify gas saturation developed in aquifer prior to the 1978 Izu-Oshima-Kinkai earthquake. Henry's coefficients (H) for radon at aquifer temperature (14 °C) is 3.27 (Weigel 1978; Rogers 1958). Based on the anomaly data in Fig. 7.2a, groundwater-radon concentration declined from a background level of 483 ± 3 cpm to a minimum level of 439 ± 7 cpm. Applying Eq. (7.1), the gas saturation developed prior to the 1978 Izu-Oshima-Kinkai earthquake can be determined. A gas saturation of 3.07% developed in aquifer prior to the 1978 Izu-Oshima-Kinkai earthquake (Tsunomori and Kuo 2010).

In a low-porosity fractured aquifer under undrained conditions and compressive stress, the volumetric strain is a linear function of the gas saturation. The fracture porosity is the slope of the straight line (Kuo et al. 2011; Kuo and Tsunomori 2014).

$$de \cong \phi S_g \tag{7.2}$$

where de is volumetric strain, dimensionless; ϕ is initial fracture porosity before rock dilatancy, fraction; S_g is gas saturation, %. With the help of Eq. (7.1), radon anomaly can be used to calculate gas saturation (S_g). Given the fracture porosity (ϕ), volumetric strain (de) can be estimated for a low-porosity confined aquifer with the help of Eq. (7.2). Vice versa, given the volumetric strain (de), fracture porosity (ϕ) can be estimated for a low-porosity confined aquifer with the help of Eq. (7.2).

A precursory strain change of about 1 ppm was also measured by the Sacks-Evertson borehole strainmeter at Irozaki located 50 km from the epicenter of the 1978 Izu-Oshima-Kinkai earthquake (Furuya and Fukudome 1986). Figures 7.2a and b show precursory changes of groundwater radon and volumetric strain for the 1978 Izu-Oshima-Kinkai earthquake, respectively (Wakita 1996; Furuya and Fukudome 1986). The change of volumetric strain with time is somewhat like the precursory phenomena expected from the dilatancy-fluid diffusion model (Nur 1972; Scholz et al. 1973; Furuya and Fukudome 1986). The precursory strain change measured at Irozaki corroborates the radon anomaly precursory to the 1978 Izu-Oshima-Kinkai earthquake.

Precursory to the 1978 Izu-Oshima-Kinkai earthquake, a gas saturation of 3.07% developed in the aquifer near the SKE-1 well (Tsunomori and Kuo 2010). Given a precursory strain change of about 1 ppm measured at Irozaki, with the help of Eq. (7.2), a fracture porosity (ϕ) of 0.0000426 can be estimated near the SKE-1 well (Kuo and Tsunomori 2014).

7.3 Iceland: Radon Anomaly Before the 1978 Southern Iceland Earthquake (M_L 4.3)

The 1978 M_L 4.3 Southern Iceland earthquake occurred on November 19 in the Southern Iceland Seismic Zone (SISZ). Figure 7.3 shows the general tectonic features of Iceland and the location of the SISZ, which extends about 70 km east–west and 10 km north–south. The 50 km long rectangle of the SISZ is called the southern lowlands. Geothermal wells exist throughout the southern lowlands. Geothermal wells can be monitored for radon content. Figure 7.4 shows the southern lowlands, which is bounded by the Western Volcanic Zone on the west and by the Hreppar anticline on the east. Figure 7.4 shows the epicenter of the 1978 M_L 4.3 Southern Iceland earthquake (63N59.7 and 20W27.8). Anomalous decrease in groundwater-radon concentration prior to the 1978 M_L 4.3 Southern Iceland earthquake was observed at the radon-monitoring station Fludir (FL-W5 well) on the southern lowlands.

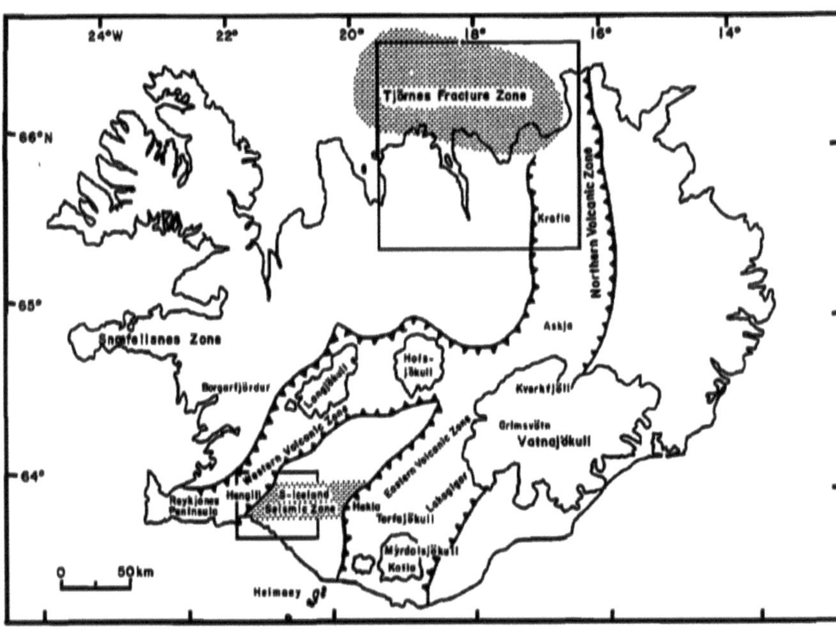

Fig. 7.3 Map of Iceland showing the general tectonic features of Iceland and the location of Southern Iceland Seismic Zone (SISZ) (Adapted from Hauksson and Goddard [1981])

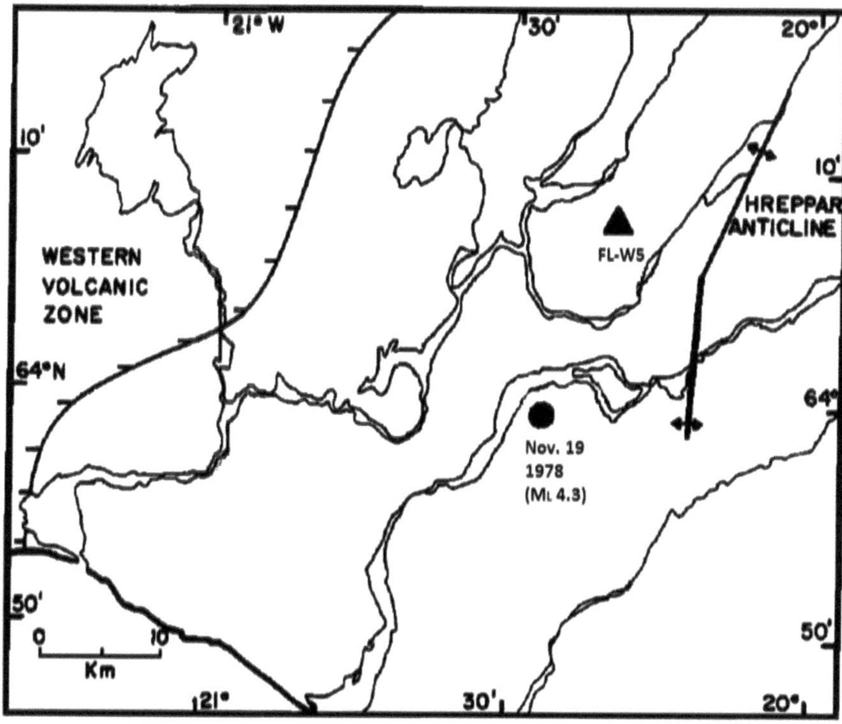

Fig. 7.4 Map showing the epicenter of the 1978 ML 4.3 Southern Iceland earthquake (63N59.7 and 20W27.8) and the radon-monitoring station Fludir (FL-W5 well) on the southern lowlands (Adapted from Hauksson and Goddard [1981])

Hauksson and Goddard (1981) reported radon anomaly precursory to the 1978 M_L 4.3 Southern Iceland earthquake. Figure 7.5 shows the radon anomaly observed at the radon sampling station Fludir (FL-W5 well). The distance between the epicenter and the radon sampling station Fludir (FL-W5 well) was about 16 km for the 1978 M_L 4.3 Southern Iceland earthquake. At the beginning of November 1978, the radon concentration at station Fludir (FL-W5 well) began to decrease from a background level of 533 dpm/kg H_2O eighteen days before the 1978 M_L 4.3 Southern Iceland earthquake (Fig. 7.5). In mid-November 1978 the radon concentration reached a minimum level of 107 dpm/kg H_2O. Then the concentration reversed and began to increase in seven days before the main shock, reaching the previous background level. The physical mechanism was poorly understood regarding the radon anomaly precursory to the 1978 M_L 4.3 Southern Iceland earthquake (Hauksson and Goddard 1981).

The radon anomaly precursory to the 1978 M_L 4.3 Southern Iceland earthquake observed at FL-W5 well in the Southern Iceland Seismic Zone (Hauksson and Goddard 1981) can be compared with that observed at Antung D1 well precursory to the 2003 M_w 6.8 Chengkung earthquake (Kuo et al. 2006). The 20-km distance

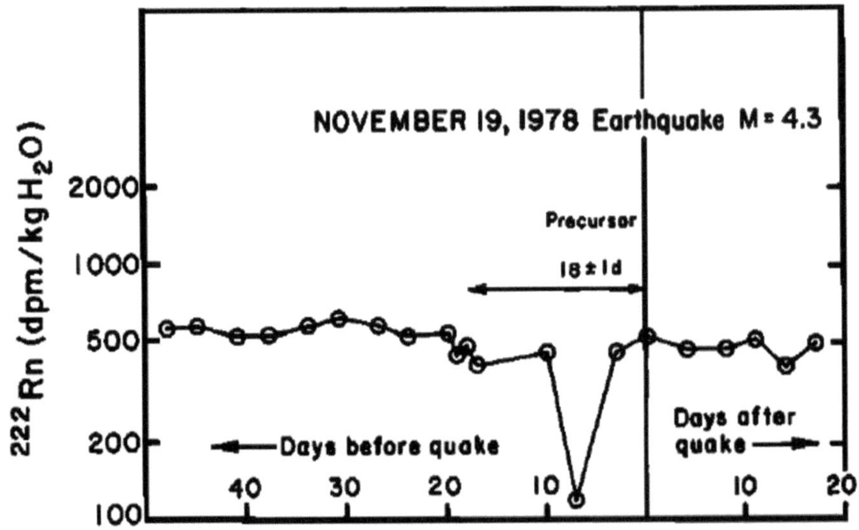

Fig. 7.5 Radon anomaly observed at the radon-monitoring station Fludir (FL-W5 well) (Adapted from Hauksson and Goddard [1981])

from the epicenter to the radon-monitoring station (Antung D1) for the 2003 M_w 6.8 Chengkung earthquake is close to the 16 km distance from the epicenter to station distance for the 1978 M_L 4.3 Southern Iceland earthquake. However, the precursory time is significantly shorter before the 1978 M_L 4.3 Southern Iceland earthquake than that before the 2003 M_w 6.8 Chengkung earthquake. At a precursory time of sixty-five days before the 2003 Chengkung earthquake, the radon concentration of groundwater started to decrease. The radon concentration started to decrease at a precursory time of only eighteen days before the 1978 Southern Iceland earthquake. The 1978 M_L 4.3 Southern Iceland earthquake occurred on a divergent plate boundary. The 2003 M_w 6.8 Chengkung earthquake occurred on a convergent plate boundary. Precursory time observed at a divergent plate boundary appears shorter than that observed at a convergent plate boundary.

Discrete samples of geothermal water were collected from the radon-monitoring well (FL-W5) for analysis of radon content. The wellhead temperature is 94 °C. The well is artesian with a total depth of 321 m. The sampling frequency was about twice per week. A sampling method was developed to draw a total sample suitable for the geothermal waters of two-phase system. An evacuated bottle was connected to the wellhead and a vacuum gauge. A total sample was then taken by filling the bottle until the gauge showed 1 atm pressure. In the laboratory, the radon was stripped from the total sample using helium as a carrier gas through the sampling bottle. The stripped radon was adsorbed onto activated charcoal at −60 °C. The charcoal column was evacuated to remove impurities at both −60 °C and room temperature. The charcoal column was then heated to 470 °C to liberate radon. Helium was used to transfer the liberated radon into an evacuated ZnS scintillation cell (Lucas cell). The cell was placed against photomultiplier tube in a light-tight chamber for counting.

The radon-monitoring well (FL-W5) is a geothermal well completed in fractured volcanic rocks. FL-W5 well is located 10 km west of the Hreppar anticline in the Southern Iceland Seismic Zone (SISZ). Under such geological conditions and when regional stress increases, tension fractures develop in aquifer rock at a rate faster than the rate at which pore water can recharge into the newly created fractures. The aquifer with weak recharge is often called under undrained condition. Gas saturation and two phases (vapor and liquid) could develop in the rock fractures. Meanwhile, the radon in groundwater volatilizes and partitions into the gas phase and the concentration of radon in groundwater decreases (Kuo et al. 2006). Thus, the anomalous decline in groundwater-radon concentration prior to the 1978 M_L 4.3 Southern Iceland earthquake can be interpreted.

The physical mechanism of in-situ groundwater-radon volatilization can be applied here to explain the phenomena of anomalous decrease in groundwater radon observed at FL-W5 well by Hauksson and Goddard (1981). In-situ radon-volatilization model, Eq. (7.1), correlates the observed decline in groundwater radon with the gas saturation developed in the confined aquifer (Kuo et al. 2006).

$$C_0 = C_w(H \times S_g + 1) \tag{7.1}$$

where C_0 is initial radon concentration in groundwater precursory to radon decrease, pCi/L; C_w is the minimum radon concentration in groundwater observed during radon anomaly, pCi/L; S_g is gas saturation, fraction; H is Henry's coefficient for radon at aquifer temperature, dimensionless. The observed decline in groundwater-radon concentration can be applied to quantify the gas saturation developed in the aquifer using Eq. (7.1).

Based on Weigel equation (Weigel 1978) and Rogers (1958), the Henry's coefficients (H) at FL-W5 well temperature (94 °C) is 9.21 for radon. Using the data of radon background and minima from Fig. 7.5 and Eq. (7.1), the gas saturation can be estimated prior to the 1978 M_L 4.3 Southern Iceland earthquake. Hauksson (1981) reported 80% decline of radon concentration precursory to the 1978 M_L 4.3 Southern Iceland earthquake. The radon concentration at FL-W5 well decreased from a background level of 533 dpm/kg H_2O to a minimum concentration of about 107 dpm/kg H_2O (Hauksson and Goddard 1981). It is estimated that a gas saturation of 43.2% developed precursory to the 1978 M_L 4.3 Southern Iceland earthquake. The gas saturation of 43.2% developed precursory to the 1978 M_L 4.3 Southern Iceland earthquake on the divergent plate boundary is higher than the gas saturation of 17.7% developed precursory to the 2003 M_w 6.8 Chengkung earthquake on the convergent boundary.

7.4 Taiwan: Correlations for Earthquake Prediction on the Longitudinal Valley Fault

A long-term observation of groundwater radon was initiated in July 2003 at well D1 located at the Antung hot spring. The Antung hot spring is a small low-porosity fractured aquifer situated in an andesitic block and surrounded by a ductile mudstone of the Lichi mélange (Chen and Wang 1996). Figure 7.6 shows that well D1 is 3 km southeast of the Chihshang fault (Longitudinal Valley Fault). The fault ruptured during two 1951 earthquakes of magnitudes M 6.2 and M 7.0 (Hsu 1962). The 2003 M_w 6.8 Chengkung was the strongest earthquake near the Chengkung area in eastern Taiwan since 1951. The Longitudinal Valley Fault is part of the boundary of the present-day plate suture between the Eurasia and the Philippine Sea plates.

Between July 2003 and February 2018, recurrent radon anomalies were observed at Antung well D1 to precede the seven main earthquakes shown in Fig. 7.6. These seven earthquakes are the Chengkung M_w 6.8 (December 10, 2003); Taitung M_w

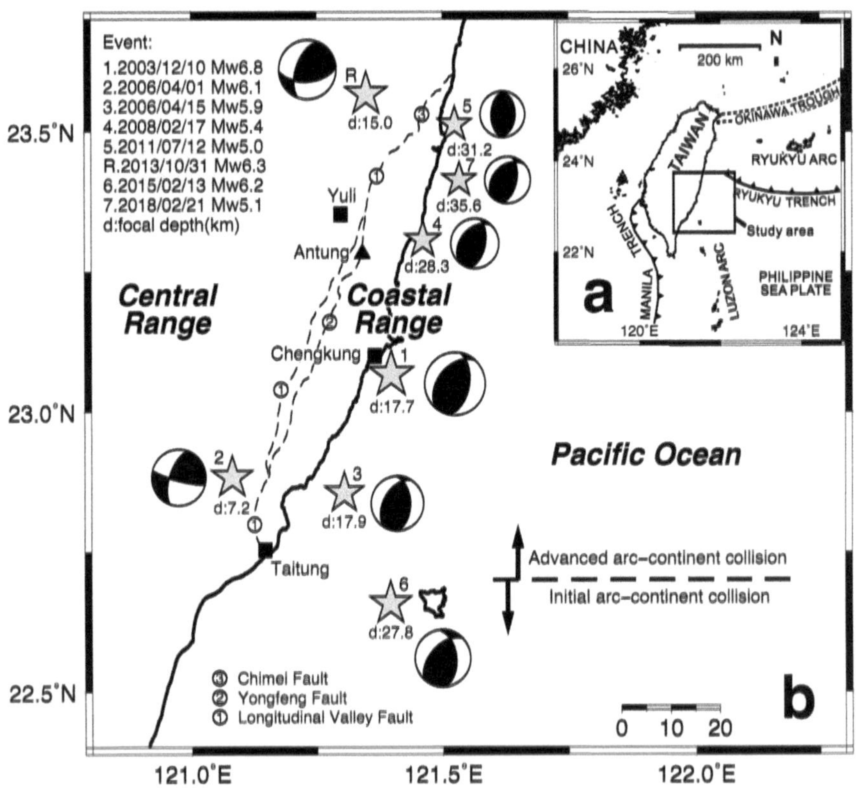

Fig. 7.6 Map of the epicenters of the large earthquakes that occurred near Antung from 2003 to 2018. **a** Map of Taiwan. **b** Study area near the Antung hot spring (filled stars: mainshocks, filled triangle: radon-monitoring well D1) (From Kuo et al. 2020)

6.1 (April 1, 2006) and M_w 5.9 (April 15, 2006); Antung M_w 5.4 (February 17, 2008); Chimei M_w 5.0 (July 12, 2011); Green Island M_w 6.2 (February 13, 2015); and Changbin M_w 5.1 (February 21, 2018) quake. Well D1 is located 24 km, 52 km, 47 km, 13 km, 32 km, 69 km, and 25 km, respectively, from the epicenters of the above seven earthquakes. The 2006 M_w 5.9 Taitung earthquake that occurred on April 15, 2006, was triggered by the 2006 M_w 6.1 Taitung earthquake that occurred on April 1, 2006 (Wu et al. 2006).

Figure 7.6 shows the epicenters and focal mechanisms of the above seven earthquakes that occurred near Antung between July 2003 and February 2018. Well D1 and the epicenters of Events 1, 2, 3, 4, 5, and 7 (the 2003 Chengkung, 2006 Taitung (two quakes), 2008 Antung, 2011 Chimei, and 2018 Changbin earthquakes) are located near the plate boundary (the Longitudinal Valley Fault), representing an advanced arc-continent collision stage of tectonic development. Figure 7.7b is a block diagram illustrating the tectonic setting near the Coastal Range currently in the stage of advanced arc-continent collision (Chen 2009). In contrast, the epicenter of offshore Event 6 (the 2015 Green Island earthquake) was located on the Luzon volcanic arc that is undergoing initial arc-continent collision. Figure 7.7a is a block diagram showing the tectonic setting near Green Island currently in the stage of initial arc-continent collision (Chen 2009). The Luzon volcanic arc extends from eastern Taiwan to the Philippines. As shown in Fig. 7.6, the boundary between the initial arc-continent collision and advanced arc-continent collision zones is currently about latitude 22.7°N (Huang et al. 2006).

The observed radon anomalies precursory to the above seven main earthquakes are exhibited in Fig. 7.8a–f. Radon concentration of well D1 at Antung decreased from background levels of 787 ± 42, 762 ± 57, 700 ± 57, 752 ± 24, 763 ± 21, and 718 ± 32 pCi/L to precursory minima of 326 ± 9, 371 ± 9, 480 ± 43, 447 ± 18, 535 ± 28, and 500 ± 17 pCi/L, respectively, prior to the 2003 Chengkung, 2006 Taitung (two quakes), 2008 Antung, 2011 Chimei, 2015 Green Island, and 2018 Changbin earthquakes. The observed radon decline increases as the earthquake magnitude increases. Figure 7.8a–f shows that all the seven radon anomalies follow a pattern that can be characterized into three stages. The rule to define Stages 1, 2, and 3 is based on the trend regarding the temporal behavior of radon concentration. During Stage 1, radon concentration is fairly stable at a background level. During Stage 2, radon concentration decreases from the background level to a minimum. During Stage 3, radon concentration increases from the minimum to the previous background level. The anomalous radon declines exhibited in Fig. 7.8a–f can be explained based on the mechanism of in-situ radon volatilization (Kuo et al. 2006).

A radon anomaly at the Antung well D1 is defined as a significant deviation from the mean value, or, three standard deviations below the mean value. As shown in Fig. 7.8a–f, the mean radon concentration and associated standard deviation are 787 ± 42, 762 ± 57, 700 ± 57, 752 ± 24, 763 ± 21, and 718 ± 32 pCi/L for the 2003 Chengkung, 2006 Taitung (two quakes), 2008 Antung, 2011 Chimei, 2015 Green Island, and 2018 Changbin earthquakes, respectively. The green rectangles in Fig. 7.8a–f are used to identify radon anomalies at the Antung well D1, which

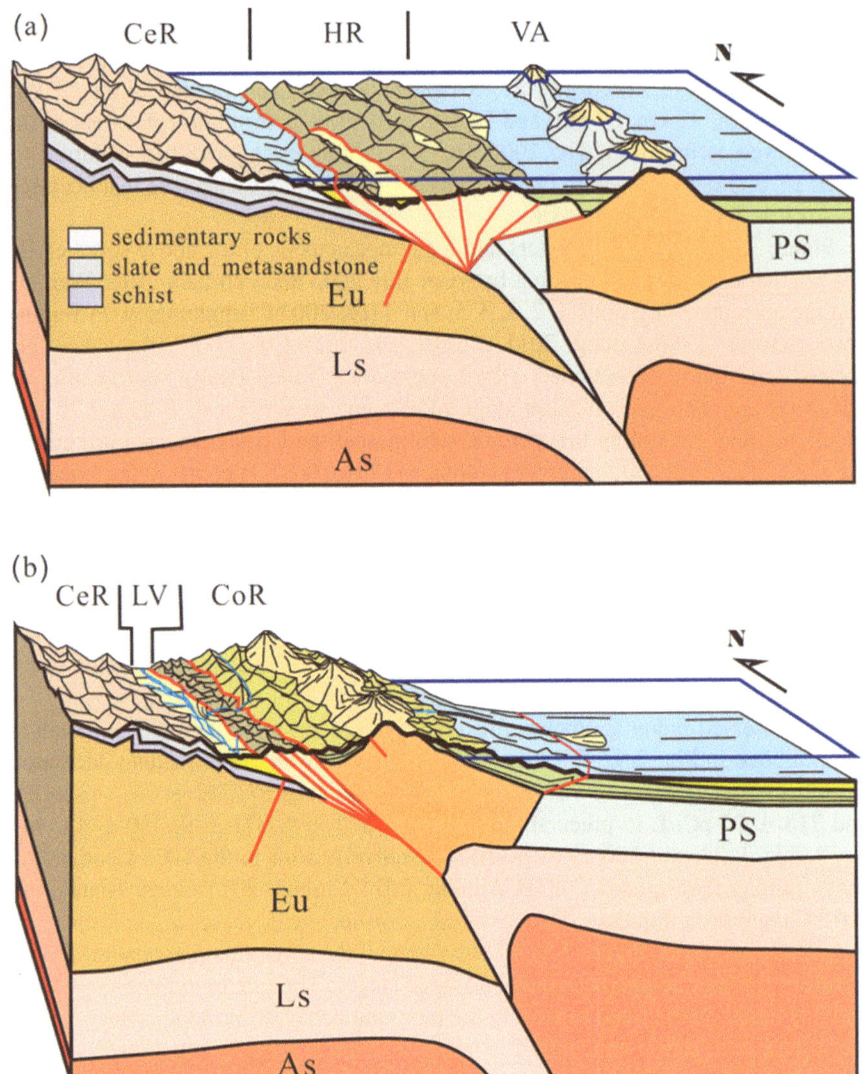

Fig. 7.7 A block diagram of tectonic framework of the SE Taiwan Offshore (no scale; Chen 2009 with publisher's permission). **a** Tectonic setting near Green Island in the stage of initial arc-continent collision (about latitude 21.0–22.7°N). **b** Tectonic setting near Coastal Range in the stage of advanced arc-continent collision (about latitude 22.7–23.5°N). As: asthenosphere; CeR: Central Range; CoR: Coastal Range; Eu: Eurasian Plate; HR: Huatung Ridge; Ls: lithosphere (upper mantle); LV: Longitudinal Valley; PS: Philippine Sea Plate; VA: North Luzon Arc (Green and Lanyu islands) (From Kuo et al. 2020)

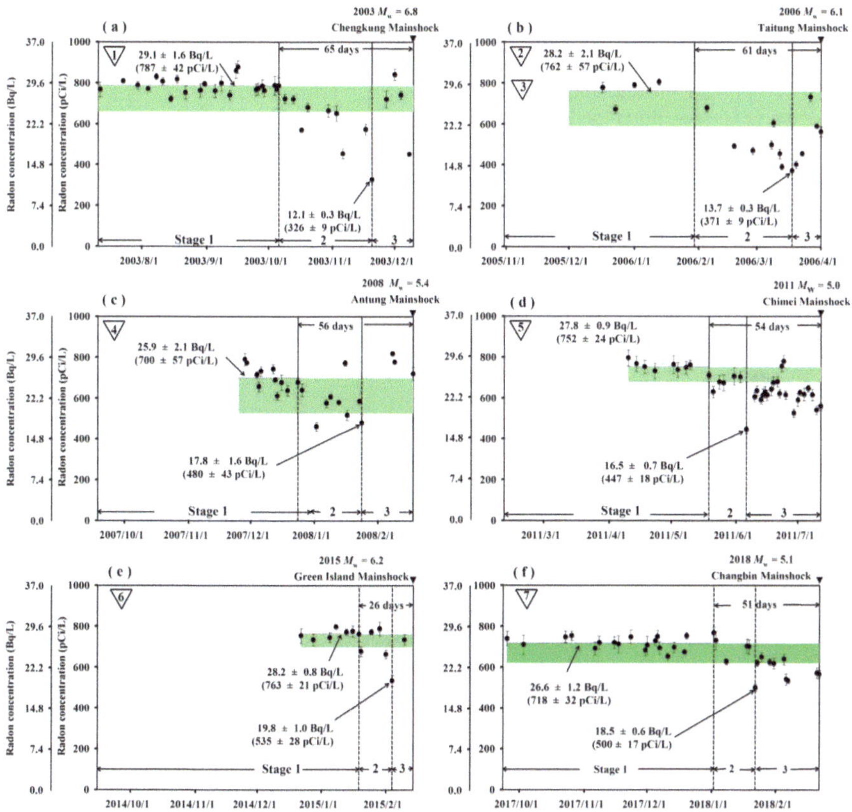

Fig. 7.8 Radon-concentration data at Antung well D1 prior to **a** 2003 Chengkung, **b** 2006 April 1 and April 15 Taitung, **c** 2008 Antung **d** 2011 Chimei, **e** 2015 Green Island, and **f** 2018 Changbin earthquakes. Green rectangles show radon concentration between the mean radon concentration and three standard deviations below the mean. Stages 1, 2, and 3 are defined in text. Numbers in inverted triangles correspond to earthquake events in Fig. 7.1 (From Kuo et al. 2020)

show radon concentration between the mean radon concentration and three standard deviations below the mean.

As shown in Fig. 7.9, Events 1, 3, 4, 5, and 7 are all thrust-type earthquakes occurring on the Longitudinal Valley Fault in advanced arc-continental collision zone. Using the radon minima precursory to the above events, the observed dimensionless radon-decline, or, $\left(\frac{C_0}{C_w} - 1\right)$ is correlated with earthquake magnitude as follows (Kuo 2014).

$$\left(\frac{C_0}{C_w} - 1\right) = 0.5118M_w - 2.0981 \tag{7.3}$$

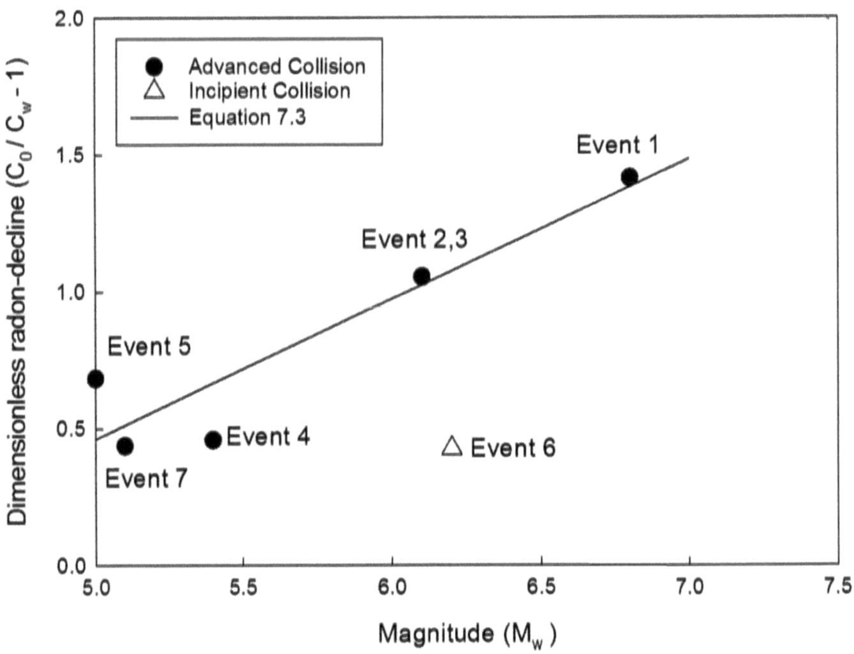

Fig. 7.9 Dimensionless radon decline observed at Antung well D1 as a function of earthquake magnitude (M_w). Event 3 was triggered by Event 2 (From Kuo et al. 2020)

where C_0 is the initial radon concentration in groundwater precursory to each radon anomaly, pCi/L; C_w is the radon minimum in groundwater observed in well D1 during an anomalous decline, pCi/L; M_w is the earthquake magnitude.

For earthquakes occurring on a given fault (Longitudinal Valley Fault), the observed radon minima can be correlated with earthquake magnitude and crust strain. The observed precursory minimum in radon concentration decreases as the earthquake magnitude increases. Using the radon background and minimum observed in well D1 during an anomalous decline, Eq. (7.3) has been proven useful to predict earthquake magnitude occurring on the Longitudinal Valley Fault. A long-term observation of groundwater radon is required and worthwhile to obtain such a correlation (Kuo 2014).

To predict earthquakes occurring on a given fault (Longitudinal Valley Fault), both earthquake magnitude and precursory time are required as soon as a radon anomaly is noticed. The precursor time for a radon anomaly is defined as the length of time between the moment when the concentration of groundwater radon starts to decline and the time of occurrence of the earthquake. The precursor times were 65, 61, 56, 54, and 51 days for radon anomalies precursory to Event 1, 3, 4, 5, and 7, respectively. The observed precursory time increases as the earthquake magnitude increases for large main earthquakes occurring on the Longitudinal Valley Fault. The observed precursor time of radon anomaly is correlated with earthquake magnitude as follows.

$$Log_{10}T = 0.0530M_w + 1.4561 \quad (7.4)$$

where T is the precursor time of a radon anomaly, day; M_w is the earthquake magnitude. Equations (7.3) and (7.4) provide a quantitative means to forecast both magnitude and precursory time for large earthquakes occurring on the Longitudinal Valley Fault. As soon as a radon minimum is observed at well D1, we can make an early warning of magnitude and precursory time for a coming earthquake on the Longitudinal Valley Fault using Eqs. (7.3) and (7.4).

Event 6 (the 2015 M_w 6.2 Green Island earthquake) did not occur on the Longitudinal Valley Fault. Here, we can investigate the effect of causative fault on precursory behavior of groundwater radon. The observed radon anomaly prior to the 2015 Green Island quake (Event 6) is shown in Fig. 7.8e. Notice that the precursory time (26 days) observed prior to Event 6 is significantly smaller than those observed prior to Events 1, 3, 4, 5, and 7 (65, 61, 56, 54, and 51 days). The open triangle shown in Figs. 7.9 and 7.10 represents Event 6 (the 2015 Green Island earthquake). The solid circles shown in Figs. 7.9 and 7.10 are the earthquakes occurring on the Longitudinal Valley Fault (Events 1, 3, 4, 5, and 7) and can be correlated as a regressed line with Eqs. (7.3) and (7.4), respectively. As shown in Fig. 7.9, the observed radon decline prior to Event 6 is considerably smaller than those observed prior to Events 1, 3, 4, 5, and 7. Figure 7.10 also shows that the observed precursory time prior to Event 6 is significantly smaller than those observed prior to Event 1, 3, 4, 5, and 7. Figures 7.9 and 7.10 imply that the correlations between radon decline, precursory time, and earthquake magnitude are characteristics of the causative fault, which plays an important role in stress transfer. Given radon-monitoring well D1 located in advanced arc-continental collision zone and the epicenter of Event 6 located in initial arc-continent collision zone, it takes additional time lag and attenuation for the stress transfer from one tectonic setting to the other.

Anomalous decreases in the concentration of groundwater radon were observed prior to seven main earthquakes, Events 1–7 (the 2003 M_w 6.8 Chengkung, 2006 M_w 6.1 and M_w 5.9 Taitung, 2008 M_w 5.4 Antung, 2011 M_w 5.0 Chimei, 2015 M_w 6.2 Green Island, and 2018 M_w 5.1 Changbin earthquakes). However, no anomalous decline in the concentration of groundwater radon was detected at well D1 prior to the 2013 Rueisuei M_w 6.3 earthquake, Event R (Kuo et al. 2017). Figure 7.6 also shows the epicenter location and strike-slip focal mechanism of 2013 Rueisuei M_w 6.3 earthquake. The 2013 Rueisuei earthquake did not occur on the Longitudinal Valley Fault. Instead, the 2013 Rueisuei earthquake occurred on the Central Range Fault. Again, the causative fault affects the correlations between radon decline, precursory time, and earthquake magnitude.

Unlike Events 1–5, why was no anomalous decline in the concentration of groundwater radon detected at well D1 prior to Event R (Kuo et al. 2017)? The detectability of anomalous declines in groundwater radon may depend on the stress conditions in aquifer rock. It is fortunate to have a nearby seismic station (HWA055) for understanding the stress conditions in aquifer rock near the Antung hot spring area during

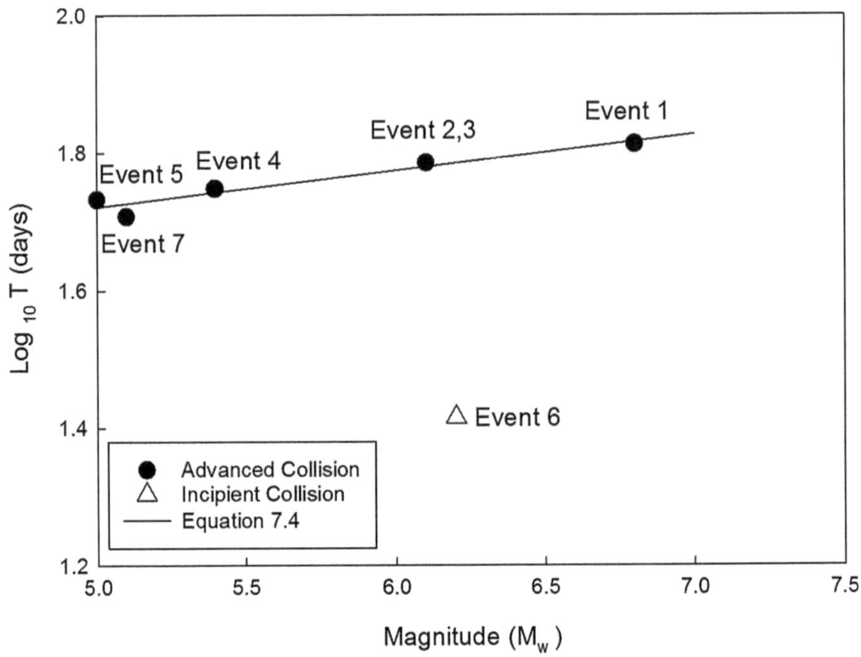

Fig. 7.10 Precursor time of radon anomaly observed at Antung well D1 as a function of earthquake magnitude (M_w). Event 3 was triggered by Event 2 (From Kuo et al. 2020)

earthquakes (Events 1–s5 and Event R). The distance between the seismic station and Antung well D1 is 2.29 km. Figure 7.11 shows an upward movement of the P wave first motion for Events 1–5 with detectable anomalous declines in groundwater radon. In contrast, for Event R without detectable anomalous declines in groundwater radon, the P wave first motion is downward. The first motion of the P-waves indicates the stress condition in rocks, compression or tension, at the time of the earthquake. Figure 7.11 suggests a possible relationship between the stress conditions in aquifer rock and the detectability of precursory radon declines. For aquifer rock under compressive conditions before and at the earthquake, a precursory anomalous decrease in the concentration of groundwater radon can be detected (Events 1–5). For aquifer rock under tension conditions before and at the earthquake (Event R), no precursory change in the concentration of groundwater radon was observed.

7.5 Future Prospects

Recurrent groundwater-radon anomalies have been observed at the Antung hot spring (well D1) prior to the main local earthquakes in the Taiwan subduction zone. A small fractured aquifer can be used as an effective natural strainmeter to catch precursory

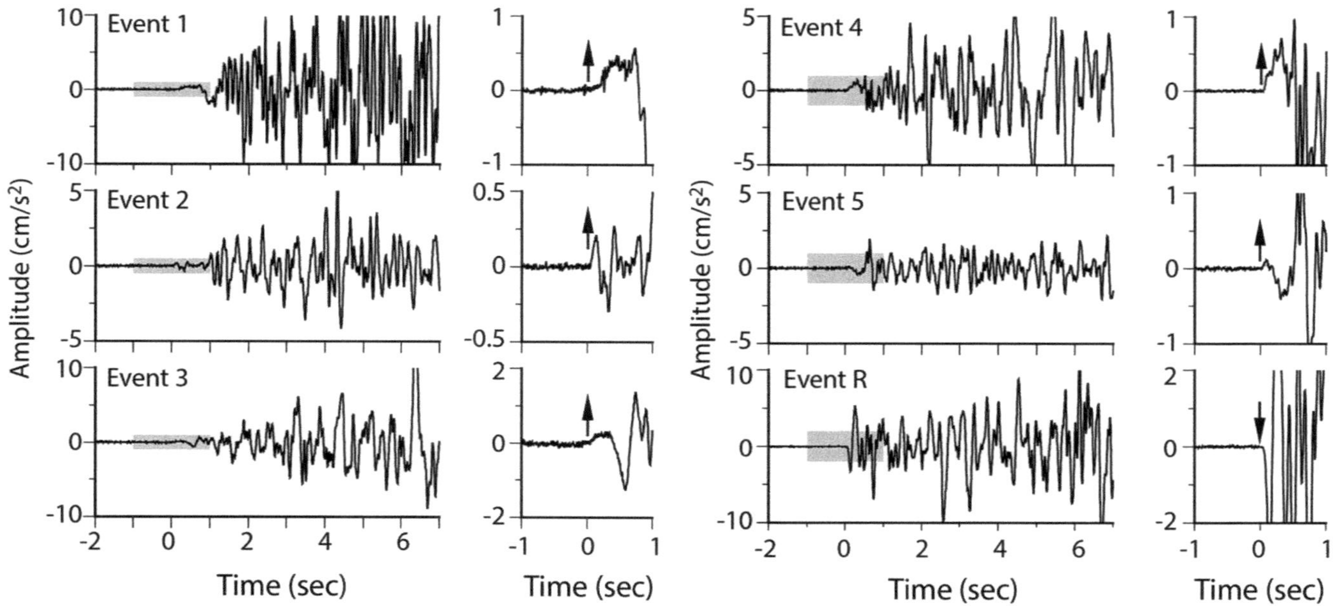

Fig. 7.11 Seismic waveforms of Events 1–5 and R recorded in HWA055 station. Gray rectangular area: the zoomed region of the seismic waveform of the P wave polarity on the right panel for each event. Black arrow: P wave first motion (From Kuo et al. 2017)

declines in groundwater radon for earthquake warnings. A long-term monitoring of groundwater radon can provide useful data to establish the correlations between radon decline, precursory time, and earthquake magnitude for an active fault.

Faults play important roles in stress transfer and affect the radon anomalies precursory to local large earthquakes. The correlations between radon decline, precursory time, and earthquake magnitude are characteristics of the causative fault. The recurrence of a large earthquake on a given active fault takes years. A long-term monitoring of groundwater at a suitable geological site is required to obtain the correlations such as Eqs. (7.3) and (7.4), which are useful for early warning of local disastrous earthquakes.

Via a basic observation of groundwater radon at Antung well D1, all large thrust-type main earthquakes on the Longitudinal Valley Fault can be warned months in advance. The findings in this book have significant merit on a local or regional basis and most importantly, they can perhaps be applied to other areas of the world with similar tectonic settings and physical–chemical relationships.

Identifying precursors prior to earthquakes is a long-sought goal. Gas bubbles induce two detectable earthquake precursors, groundwater radon and aquifer transmissivity. Aquifer transmissivity can complement groundwater radon as an earthquake precursor. Simultaneous monitoring of aquifer transmissivity and groundwater radon is highly recommended for future projects in earthquake precursors.

References

Chen WS, Wang Y (1996) Geology of the coastal range, eastern Taiwan. Geology of Taiwan 7

Chen WS (2009) Tectonostratigraphic framework and age of the volcanic-arc and collision basins in the Coastal Range, eastern Taiwan. Western Pacific Earth Sciences 9:67–98

Furuya I, Fukudome A (1986) Characteristics of borehole volume strainmeter and its application to seismology. J Phys Earth 34(3):257–296

Hauksson E (1981) Radon content of groundwater as an earthquake precursor: Evaluation of worldwide data and physical basis. J Geophys Res: Solid Earth 86(B10):9397–9410

Hauksson E, Goddard JG (1981) Radon earthquake precursor studies in Iceland. J Geophys Res: Solid Earth 86(B8):7037–7054

Hsu TL (1962) Recent faulting in the Longitudinal Valley of eastern Taiwan. Mem Geol Soc China 1:95–102

Huang CY, Yuan PB, Tsao SJ (2006) Temporal and spatial records of active arc-continent collision in Taiwan: a synthesis. Geol Soc Am Bull 118:274–288

Kuo MCT et al (2006) A mechanism for anomalous decline in radon precursory to an earthquake. Groundwater 44(5):642–647

Kuo T (2014) Correlating precursory declines in groundwater radon with earthquake magnitude. Groundwater 52(2):217–224

Kuo T, Tsunomori F (2014) Estimation of fracture porosity using radon as a tracer. J Petrol Sci Eng 122:700–704

Kuo T et al (2011) Correlating recurrent radon precursors with local earthquake magnitude and crust strain near the Chihshang fault of eastern Taiwan. Nat Hazards 59:861–869

Kuo T et al (2017) A stress condition in aquifer rock for detecting anomalous radon decline precursory to an earthquake. Pure Appl Geophys 174:1291–1301

Kuo T et al (2020) Precursory behavior of groundwater radon in southeastern Taiwan: effect of tectonic setting in the subduction zone. Pure Appl Geophys 177:2877–2887
Noguchi M, Wakita H (1977) A method for continuous measurement of radon in groundwater for earthquake prediction. J Geophys Res 82(8):1353–1357
Nur A (1972) Dilatancy, pore fluids, and premonitory variations of ts/tp traval times. Bull Seismol Soc Am 62(5):1217–1222
Rogers AS (1958) Physical behavior and geologic control of radon in mountain streams. USGS Bulletin 1052-E, United States Government Printing Office, Washington, DC
Scholz CH, Sykes LR, Aggarwal YP (1973) Earthquake prediction: a physical basis. Science 181(4102):803–810
Shimazaki K, Somerville P (1978) Summary of the static and dynamic parameters of the Izu-Oshima-Kinkai earthquake of January 14, 1978. Bull Earthq Res Inst Tokyo Univ 53:613–628
Tsuneishi Y (1978) Surface faulting associated with the 1978 Izu-Oshima-kinkai earthquake. Bull Earthq Res Inst Tokyo Univ 53:649–674
Tsunomori F, Kuo T (2010) A mechanism for radon decline prior to the 1978 Izu-Oshima-Kinkai earthquake in Japan. Radiat Meas 45(1):139–142
Wakita H (1996) Geochemical challenge to earthquake prediction. Proc Natl Acad Sci USA 93:3781–3786
Wakita H et al (1980) Radon anomaly: a possible precursor of the 1978 Izu-Oshima-kinkai earthquake. Science 207(4433):882–883
Weigel F (1978) Radon. Chemiker Zeitung 102(9):287–299
Wu YM et al (2006) Seismogenic structure in a tectonic suture zone: With new constraints from 2006 Mw6.1 Taitung earthquake. Geophys Res Lett 33:L22305

MIX
Papier aus verantwortungsvollen Quellen
Paper from responsible sources
FSC® C105338
FSC
www.fsc.org

If you have any concerns about our products,
you can contact us on
ProductSafety@springernature.com

In case Publisher is established outside the EU,
the EU authorized representative is:
Springer Nature Customer Service Center GmbH
Europaplatz 3, 69115 Heidelberg, Germany

Printed by Libri Plureos GmbH
in Hamburg, Germany